Introduction to Oscillatory Motion with Mathematica®

This page intentionally left blank

Introduction to Oscillatory Motion with Mathematica®

Steven Tan

Freelance Physicist

Gedanken Books

Introduction to Oscillatory Motion With Mathematica

Copyright © 2018, 2016 Steven Tan

First Edition Revised May 2018

Published by Gedanken Books

ISBN   978-1-365-38812-5

All rights reserved. Apart from any fair dealing for the purposes of research or private study or criticism or review, no part of this publication may be reproduced, stored or transmitted in any form or by any means, without the prior permission in writing of the publisher.

Email: GedankenBooks@gmail.com

Contents

Preface · vii

Chapter 1. Introduction to Mathematica · 1

1.1 Executing Simple Expressions
1.2 Defining a Function
1.3 Lists
1.4 Plotting Graphs
1.5 Calculus
1.6 Differential Equations

Chapter 2. Linear Oscillatory Systems · 23

2.1 Linear Simple Harmonic Motion
    2.1.1 Energy Considerations in Simple Harmonic Motion
    2.1.2 Representation on the Phase Plane
    2.1.3 Two-dimensional Harmonic Oscillator

2.2 Linear Oscillator in the Presence of Frictional Force
    2.2.1 Case 1. Underdamped Motion
        2.2.1.1 Energy Considerations in Underdamped Motion
        2.2.1.2 Representation on the Phase Plane
    2.2.2 Case 2. Critically Damped Motion
        2.2.2.1 Energy Considerations in Critically Damped Motion
        2.2.2.2 Representation on the Phase Plane
    2.2.3 Case 3. Overdamped Motion
        2.2.3.1 Energy Considerations in Overdamped Motion
        2.2.3.2 Representation on the Phase Plane

2.3 Sinusoidal Driven Oscillations
    2.3.1 Resonance Phenomena
    2.3.2 Energy Considerations in Resonance
    2.3.3 Representation on the Phase Plane
    2.3.4 Impulsive Forcing Functions
        2.3.4.1 Response to a Force Step Function
        2.3.4.2 Response to an Impulsive Force Function

Chapter 3. Nonlinear Oscillatory Systems · 73

3.1 Simple Nonlinear Oscillator
    3.1.1 Representation on the Phase Plane

3.2 Simple Nonlinear Pendulum
        3.2.1 Representation on the Phase Plane

3.3 Damped, Driven Nonlinear Pendulum
        3.3.1 Representation on the Phase Plane
        3.3.2 Bifurcation

3.4 The van der Pol Oscillator
        3.4.1 Non-driven van der Pol Oscillator
        3.4.2 Driven van der Pol Oscillator

Bibliography · 111

Preface

This book is a survey of basic oscillatory concepts with the aid of Mathematica® computer algebra system to represent them and to calculate with them. It is written for students, teachers, and researchers needing to understand the basic of oscillatory motion or intending to use Mathematica® to extend their knowledge. All illustrations in the book can be replicated and used to learn and discover oscillatory motion in a new and exciting way. It is meant to complement the analytical skills and to use the computer to visualize the results and to develop a deeper intuitive understanding of oscillatory motion by observing the effects of varying the parameters of the problem.

We assume the reader has some familiarity with Mathematica®, so we could focus on the physics of oscillatory motion. This book does not discuss programming in Mathematica® nor does it teach all the principles and techniques of applications using Mathematica®. However, Chapter 1 provides a basic introduction to Mathematica®, developed by Wolfram Research. New users will find that the materials in Chapter 1 enable them to become familiar with Mathematica® within a few hours. The reader can learn the essentials of Mathematica® through examples described in the book. Mathematica® commands and techniques are introduced as the need arises.

The author has used Mathematica® version 11 in the preparation of the material. All Mathematica® codes have been kept as simple as possible and should be backward compatible with earlier versions of Mathematica® or have equivalent representations. The codes should also run under later versions.

Chapter 2 deals with linear oscillatory systems which are standard discussions in most undergraduate textbooks. It ends with solutions of oscillatory systems in response to impulsive forcing functions.

A variety of numerical techniques are available in Mathematica® when solving nonlinear oscillatory systems when an analytic solution does not exist or is very difficult to find. The construction of Poincare section and the specific example of bifurcation in damped, driven, nonlinear pendulum is represented in Chapter 3, where the concept of chaos is slightly mentioned. Chapter 3 ends with an example of nonlinear system, the van der Pol oscillator.

This book is informed by the interests of the author in using Mathematica® to learn and discover physics in a new and exciting way.

The prerequisites for using this book are undergraduate courses in calculus, classical mechanics, and ordinary differential equations. Knowledge of computer programming would be beneficial but not essential.

Although extreme care was taken to correct all the misprints, it is very unlikely that I have been able to catch all of them. I shall be most grateful to those readers kind enough to bring to my attention any remaining mistakes, typographical or otherwise. Please feel free to contact me at:

GedankenBooks@gmail.com

Steven Tan

Chapter 1. Introduction to Mathematica

Initialization

This section is reserved for Mathematica notebook initialization.

At the start of each chapter, we shall clear all values and definitions of any previously defined symbols using **Clear** command. Here is also the place where we set some default options. Most Mathematica modules are defined in the texts when the needs arise.

In[1]:= `Clear["Global`*"]`

1.1 Executing Simple Expressions

In Mathematica, any text enclosed in matching **(*** and ***)** denotes comments. A comment can be inserted anywhere in Mathematica code.

To execute an expression, place the cursor at the expression, and then press [SHIFT]+[ENTER] on a PC or [COMMAND]+[RETURN] on a Mac.

Or alternatively, from the menu bar on top of Mathematica, click on the **Evaluation ⇒ Evaluate Cells**. (click on the "**Evaluation**" menu and then follows by clicking "**Evaluate Cells**")

In[2]:= `(* Addition. Type 1+2 and then execute *)`
`1 + 2`

Out[2]= 3

Parentheses, (), are used only for grouping such as the expression below

In[3]:= `(* Multiplication *)`
`(1 + 2) * 3`

Out[3]= 9

To assign the value to a variable *a*, we use **equal (=)** sign. Notice that equal sign in Mathematica is used for assigning value to a variable. It is not the same meaning as equality in standard mathematical expressions.

Since all built-in Mathematica objects begin with capital letters, they are reserved names. Mathematica will complain if we assign value to a reserved objects. Some reserved names are: **Cos, Pi, Sin, Exp**. Some function names are: **Plot**, **FindRoot**, **ParametricPlot**. The only Greek character reserved in Mathematica is π.

Some common mistakes:

In[4]:= `sin[pi]`
Out[4]= sin[pi]

In[5]:= `cos[pi]`
Out[5]= cos[pi]

The correct expressions are

In[6]:= `Sin[Pi]`
Out[6]= 0

In[7]:= `Cos[π]`
Out[7]= -1

Mathematica is case-sensitive. A variable name **A** is different from a variable name **a**. So it is a good practice to name variables begin with lowercase letters.

Let us compute the expression $(1 + 2) \times 3$ and assign the result to a variable called *a*

In[8]:= `a = (1 + 2) * 3`
Out[8]= 9

A **semicolon** (;) at the end of a Mathematica command will suppress output. For example:

In[9]:= `b = (1 + 2) * 3;`

To check the value of **b**, we type the following expression and execute it

In[10]:= (* Type b and then execute *)
b

Out[10]= 9

Multiplication in Mathematica can be represented by a **space between expressions**, for example

In[11]:= (* (1+2) and multiplied by 3 and the result is assigned
    to c. A space is used in between to denote multiplication *)
c = (1 + 2) 3

Out[11]= 9

In[12]:= (* Check if a equals b equals c *)
a == b == c

Out[12]= True

Notice that Mathematica uses **double equal sign** (==) to mean equality to check if a equals b equals c. Mathematica keeps records of all the inputs and outputs in a session. Thus, we can refer to the last result generated by using the **percent sign** (%) notation. Double percent signs (%%) gives the next-to-last result generated. %%...% (**k** times) gives the kth previous result and %k refers to output line numbered **k**.

1.2 Defining a function

Besides the built-in functions, we can also define a function in Mathematica. For example, a function f which takes an input x, and output $x^2 - 4$,

$$f(x) = x^2 - 4$$

we type in

In[13]:= f[x_] := x^2 - 4

Notice that the function argument x must be followed by an underscore (_) inside the square bracket **[ ]**. Here we have used a **colon-equal sign** (:=) to separate the left side of the function definition from the right.

To compute the output of the function when x = 1, we type in the following and execute the expression

In[14]:= **f[4]**

Out[14]= 12

Similarly, to compute the output of the function when $x = -1$

In[15]:= **f[-1]**

Out[15]= -3

We can also use Mathematica built-in function such as **Sin** in our function. For instance
$$g(x) = x - \sin(x^2).$$

In[16]:= **g[x_] := x - Sin[x^2]**

In[17]:= **g[1]**

Out[17]= 1 - Sin[1]

The result may not be what we wanted. If we wish to have numerically value, we can use the built-in function **N**. Here is an example

In[18]:= **g[2] // N**

Out[18]= 2.7568

The above expression is equivalent to

In[19]:= **N[g[2]]**

Out[19]= 2.7568

where the **N** command gives the numerical value of **g[2]**.

There are times when we need to define a different function using the same function name $f(x)$. However, $f(x)$ is still retained in memory as can be checked below

In[20]:= **? f**

```
Global`f

f[x_] := x^2 - 4
```

Thus, we want to clear out the function *f(x)* from memory using **Clear** command and then define the new function using the same function name.

In[21]:= `Clear[f]`

In[22]:= `(* Define a new function with function name f *)`
`f[x_] := 2 x^3 + 1`

Now, *f(x)* is a totally different function. We can verify this using the **?** command

In[23]:= `? f`

```
Global`f

f[x_] := 2 x^3 + 1
```

Notice in our examples in defining functions above, we have used the **:=** operator, which is called the **SetDelayed** operator. Mathematica will evaluate the expression appearing to its right afresh each time the expression appearing to its left is called.
While the **=** operator, which is called **Set** operator evaluates the expression on its right only once, at the time the assignment is made.

1.3 Lists

A list is a collection of elements separated by commas and enclosed in curly brackets. For example:
{2, 3, 5, 8}
To make a list above we can use **List** command

In[24]:= `List[2, 3, 5, 8]`
Out[24]= {2, 3, 5, 8}

We may enter a list directly without using the **List** command, such as

In[25]:= {3, 6, 7, 8}
Out[25]= {3, 6, 7, 8}

We can use lists to build up vectors, matrices, tensors and other arrays. For instance, we use **Table** command in the following to generate a one-dimensional array

In[26]:= **myList = Table$[2^i, \{i, 5\}]$**
Out[26]= {2, 4, 8, 16, 32}

and assign the list with the name **myList**.

To extract element of a list, we use **double square bracket [[and]]**. For instance, to extract the second element we type in

In[27]:= **myList[[2]]**
Out[27]= 4

To extract a sequence of elements, from the second to the fourth element of **myList**

In[28]:= **myList[[2 ;; 4]]**
Out[28]= {4, 8, 16}

To extract an element 2 from the end of a list

In[29]:= **myList[[-2]]**
Out[29]= 16

As mentioned earlier, we can build a matrix by using list. In fact, a matrix is a list of lists. For instance, we build a 3×3 matrix by typing in

In[30]:= **myMatrix = {{9, 3, 2}, {3, 3, 1}, {2, 2, 0}}**
Out[30]= {{9, 3, 2}, {3, 3, 1}, {2, 2, 0}}

To show the elements of list arranged in a regular array, we use **MatrixForm** command

In[31]:= **MatrixForm[myMatrix]**

Out[31]//MatrixForm=
$$\begin{pmatrix} 9 & 3 & 2 \\ 3 & 3 & 1 \\ 2 & 2 & 0 \end{pmatrix}$$

or equivalently

In[32]:= **myMatrix // MatrixForm**

Out[32]//MatrixForm=
$$\begin{pmatrix} 9 & 3 & 2 \\ 3 & 3 & 1 \\ 2 & 2 & 0 \end{pmatrix}$$

To switch rows and columns, we use **Transpose** command and then follow by **MatrixForm** command to show it in regular array.

In[33]:= **Transpose[myMatrix] // MatrixForm**

Out[33]//MatrixForm=
$$\begin{pmatrix} 9 & 3 & 2 \\ 3 & 3 & 2 \\ 2 & 1 & 0 \end{pmatrix}$$

The inverse of our matrix **myMatrix** is

In[34]:= **myInverse = Inverse[myMatrix]**

Out[34]= $\left\{\left\{\frac{1}{6}, -\frac{1}{3}, \frac{1}{4}\right\}, \left\{-\frac{1}{6}, \frac{1}{3}, \frac{1}{4}\right\}, \left\{0, 1, -\frac{3}{2}\right\}\right\}$

In[35]:= **myInverse // MatrixForm**

Out[35]//MatrixForm=
$$\begin{pmatrix} \frac{1}{6} & -\frac{1}{3} & \frac{1}{4} \\ -\frac{1}{6} & \frac{1}{3} & \frac{1}{4} \\ 0 & 1 & -\frac{3}{2} \end{pmatrix}$$

Let us verify that multiplication of the matrix with its inverse matrix is an identity matrix

In[36]:= **myMatrix.myInverse // MatrixForm**

Out[36]//MatrixForm=
$$\begin{pmatrix} 1 & 0 & 0 \\ 0 & 1 & 0 \\ 0 & 0 & 1 \end{pmatrix}$$

In[37]:= **myInverse.myMatrix // MatrixForm**

Out[37]//MatrixForm=
$$\begin{pmatrix} 1 & 0 & 0 \\ 0 & 1 & 0 \\ 0 & 0 & 1 \end{pmatrix}$$

1.4 Plotting Graphs

To generate a graph of a function f, we use the **Plot** command.
$$f(x) = 2x^2 + 3x - 1.$$
First of all, we clear memory of any previously defined variable f and then define the function

In[38]:= **Clear[f]**

In[39]:= **f[x_] := 2 x^2 + 3 x - 1**

In[40]:= **(* Plot the function for -2≤x≤2 *)**
Plot[f[x], {x, -2, 2}, PlotStyle → Thickness[0.005]]

Out[40]=
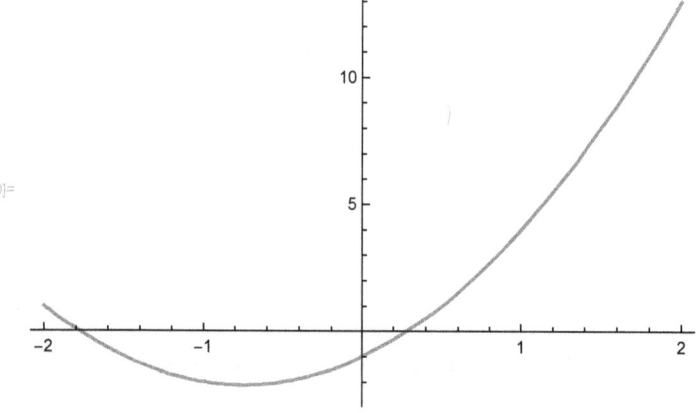

We may also write the function explicitly inside the Plot command, such as

In[41]:= `Plot[Sin[x], {x, 0, 4 π}, PlotStyle → Thickness[0.005]]`

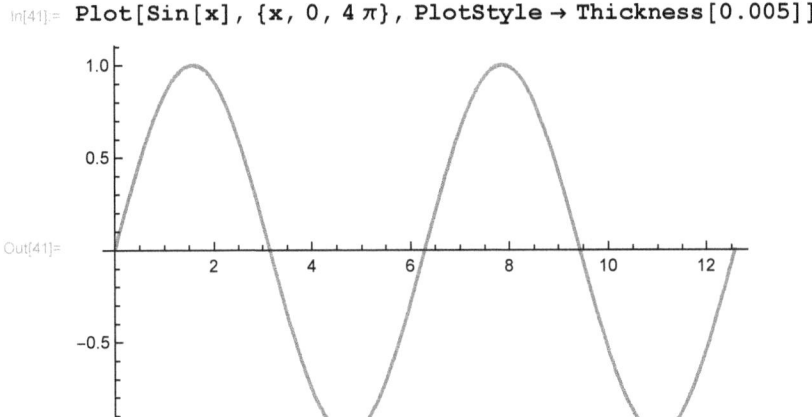

To plot several functions at once, we separate the functions with comma inside a curly bracket. The functions have different gray color using **GrayLevel** option

In[42]:= `Plot[{Sin[x], Sin[3 x], Sin[5 x]}, {x, 0, 2 π},`
 `PlotStyle → {{Thickness[0.005], GrayLevel[0.0]},`
  `{Thickness[0.005], GrayLevel[0.5]}, {Thickness[0.005], GrayLevel[0.7]}}]`

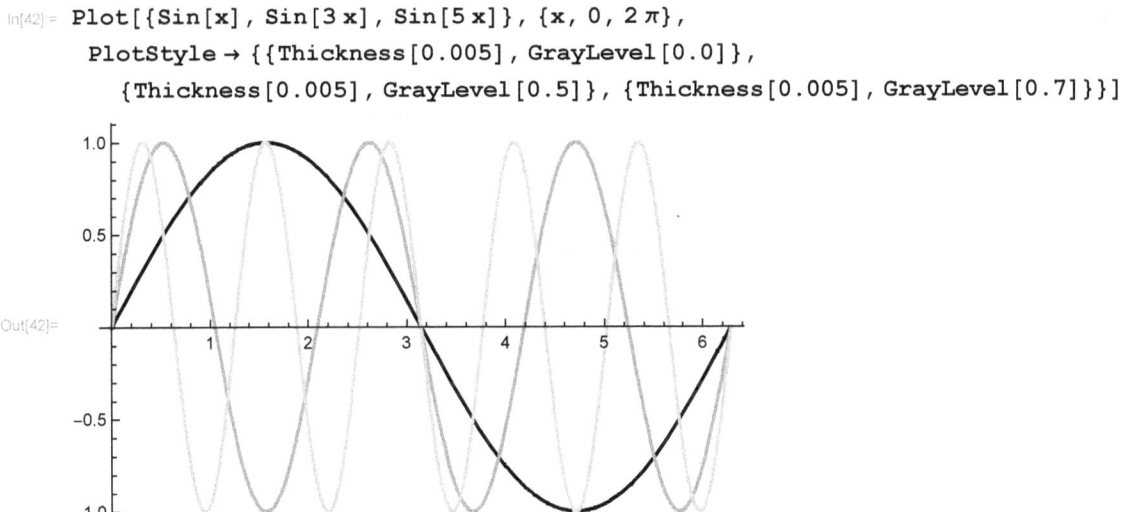

Here is another example in which we use a **Table** command to generate a list of cosine functions and sent it to **Plot** command

In[43]:= `Plot[Evaluate[Table[Cos[k x], {k, 1, 3}]], {x, 0, 2 π}, PlotStyle → Thickness[0.005]]`

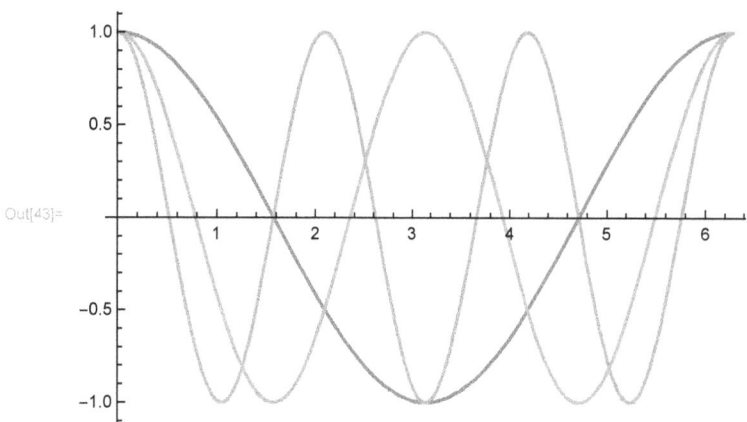

Out[43]=

Mathematica has a **Manipulate** command to allow us interactively manipulate the values of a variable of a function. For instance, we plot the function

$$\sin(k\,x),$$

in the range $0 \le x \le 2\pi$, while changing the value of k from 1 to 3.

In[44]:= `Manipulate[Plot[Sin[k x], {x, 0, 2 π}, PlotStyle → Thickness[0.005]], {k, 1, 3}]`

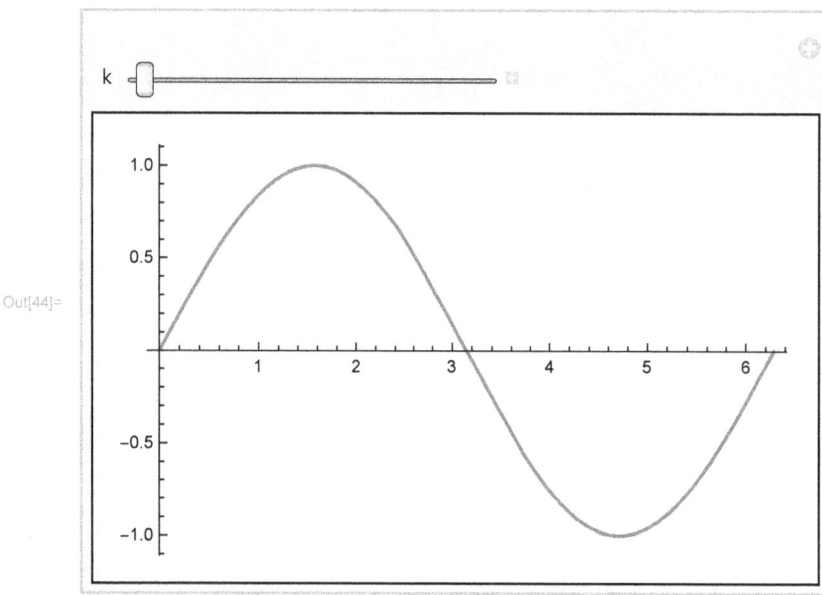

Out[44]=

Notice that **Manipulate** takes two arguments. The first is the expression to be manipulated. In this case

Plot[Sin[$k\,x$], {x, 0, 2π}].

The second is an iterator which specifies the range of values that the controller variable k is to assume. In this case k ranges from the value of 1 to 3.

Notice that there is a slider inside the frame. You can move the slider with your mouse to control the value assumed by k. Click on the ▣ button to the right of the slider to show a user control panel. Then, click on the ▶ button to watch as the curve is displayed while k changes in real time.

Here is another example of **Manipulate** command for function
$$f(x) = \sin\left(\frac{1}{x}\right),$$
in which we allow the upper limit of x to vary from p = 1 to p = 3, in steps 0.2

In[45]:= `Manipulate[Plot[Sin[1/x], {x, 0, p},`
     `PlotStyle → Thickness[0.005], PlotRange → Full], {p, 1, 3, 0.2}]`

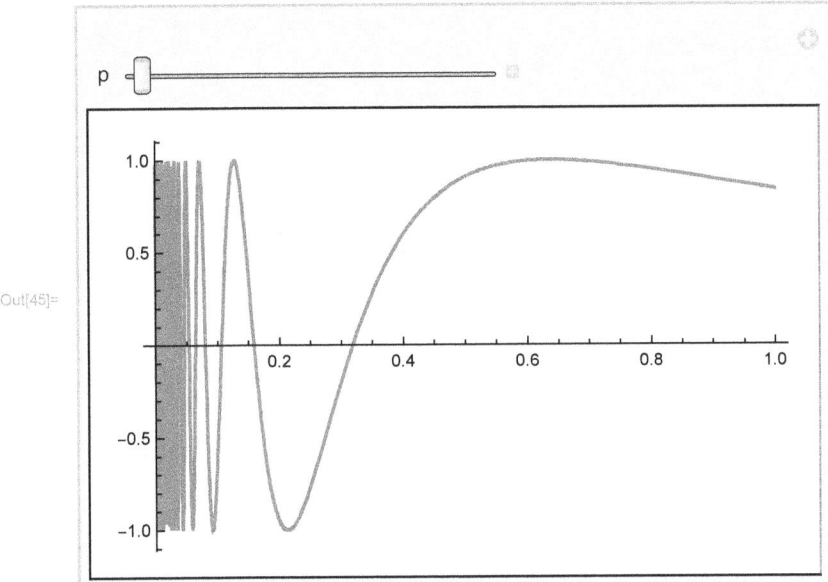

Out[45]=

Mathematica can draw curves that are defined parametrically using **ParametricPlot** command. For instance, the x- and y-coordinates of points are defined as two independent functions of a parameter t
$$\begin{cases} x(t) = \cos t \\ y(t) = \sin t \end{cases}$$

In[46]:= `Clear[x, y]`

In[47]:= `x[t_] := 3 Cos[t]`

In[48]:= `y[t_] := 2 Sin[t]`

```
In[49]:= ParametricPlot[Evaluate[{x[t], y[t]}], {t, 0, 2 π},
    PlotStyle → Thickness[0.01], AxesLabel → {"x", "y"}]
```

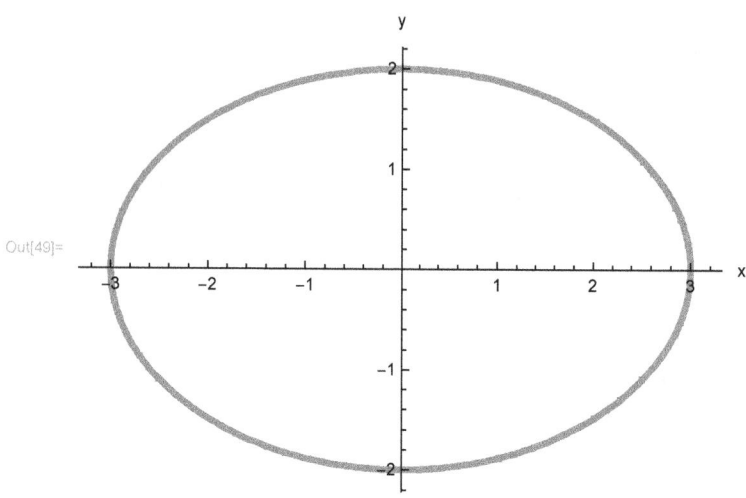

Here is another example

$$\begin{cases} x(t) = t^2 - t \\ y(t) = t^3 - 2t \end{cases}$$

In[50]:= `ParametricPlot[Evaluate[{t^2 - t, t^3 - 2 t}], {t, -2, 2},
    PlotStyle → Thickness[0.01], AxesLabel → {"x", "y"}]`

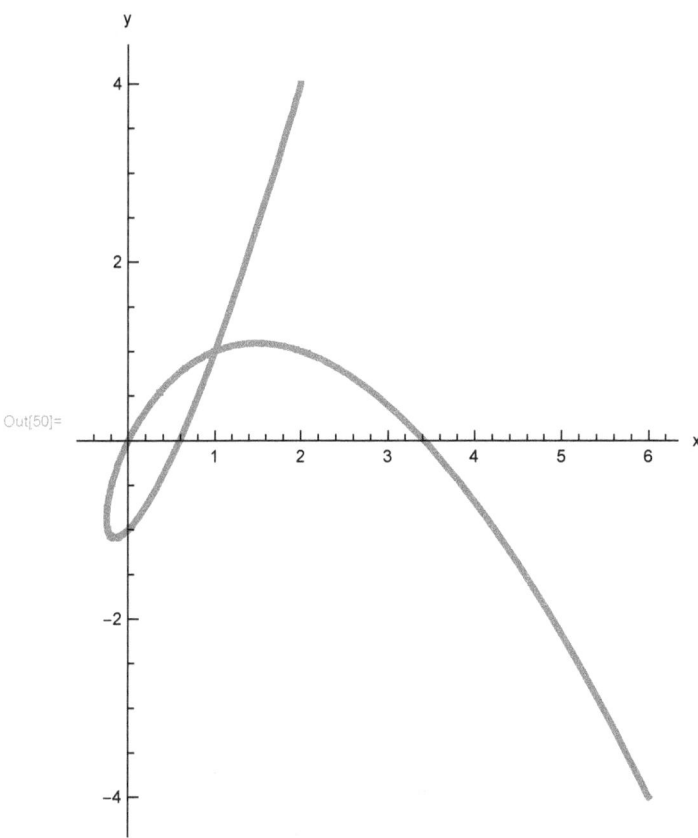

Implicitly defined curves can be graphed with the **ContourPlot** command. For example curve defined implicitly by the equation

$$x^2 + y^3 = 2 x^2 y$$

In[51]:= `ContourPlot[x² + y³ == 2 x² y, {x, -2, 2}, {y, -2, 2},`
   `Axes → True, Frame → False, ContourStyle → Thickness[0.005]]`

Out[51]=

For a function of two independent variables, the graph is a surface in three-dimensional space. Suppose a function $f(x, y)$ is defined as

$$f(x, y) = x^3 + y^3.$$

To draw the surface in three-dimensional space we can use **Plot3D** command.

In[52]:= `Plot3D[x³ + y³, {x, -2, 2}, {y, -2, 2}, AxesLabel → {"x", "y", "z"}]`

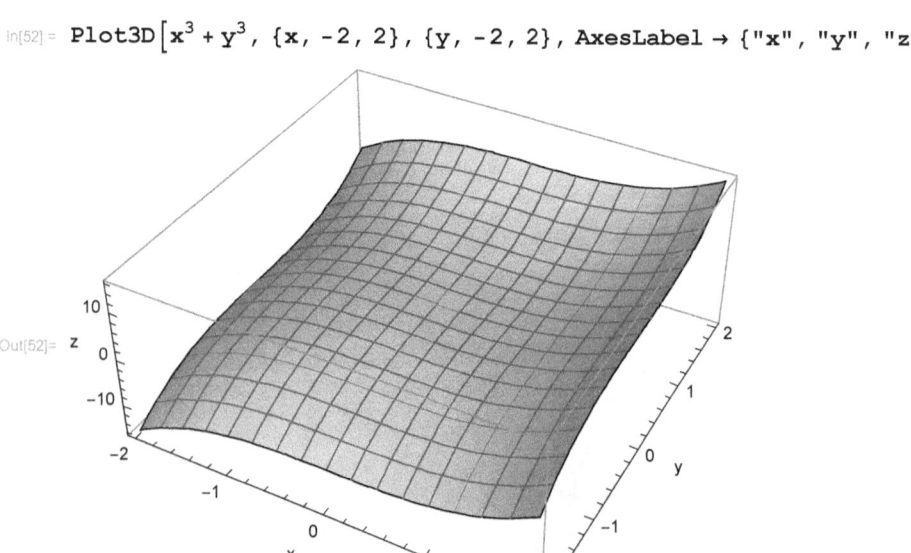

Here is another surface defined as

$$f(x, y) = \cos(x + y).$$

In[53]:= `Plot3D[Cos[x+y], {x, -π, π}, {y, -π, π}, AxesLabel → {"x", "y", "z"}]`

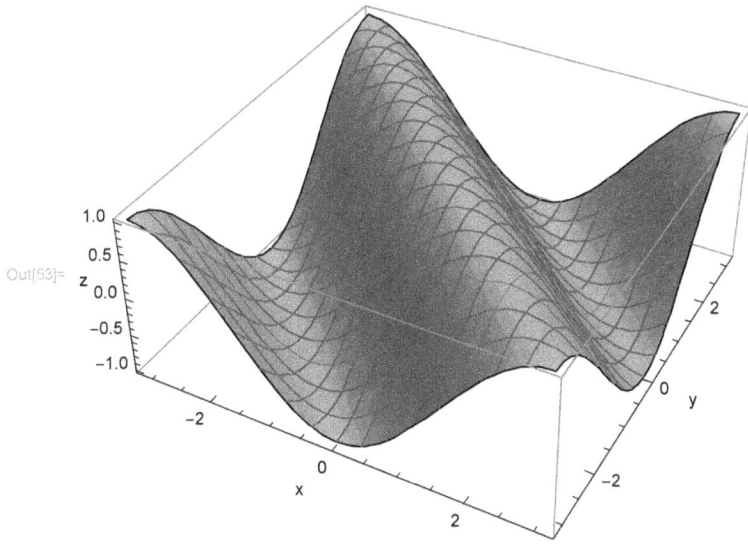

Out[53]=

If a three-dimensional space surface is defined parametrically by two parameters, s and t, such as

$$f(t) = (\cos s, \sin t, s),$$

we can use **ParametricPlot3D** to draw the surface.

In[54]:= `ParametricPlot3D[Evaluate[{Cos[s], Sin[t], s}], {s, 0, 2 π}, {t, 0, 2 π}]`

Out[54]=

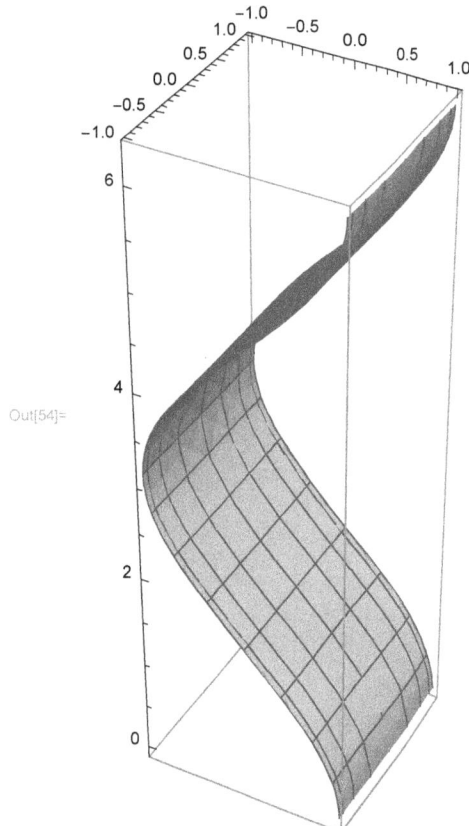

ParametricPlot3D command can also be used to produce three-dimensional space curve defined parametrically by a variable *t* only.

Suppose a vector valued function is defined as

$$\mathbf{f}(t) = (3\cos t,\ 2\sin t,\ t/2).$$

In[55]:= `Clear[f]`

In[56]:= `f[t_] := {3 Cos[t], 2 Sin[t], t/2}`

To plot the curve **f**(*t*) we type in

In[57]:= `ParametricPlot3D[Evaluate[f[t]], {t, 0, 6 π}, PlotStyle → {Red, Thickness[0.01]}]`

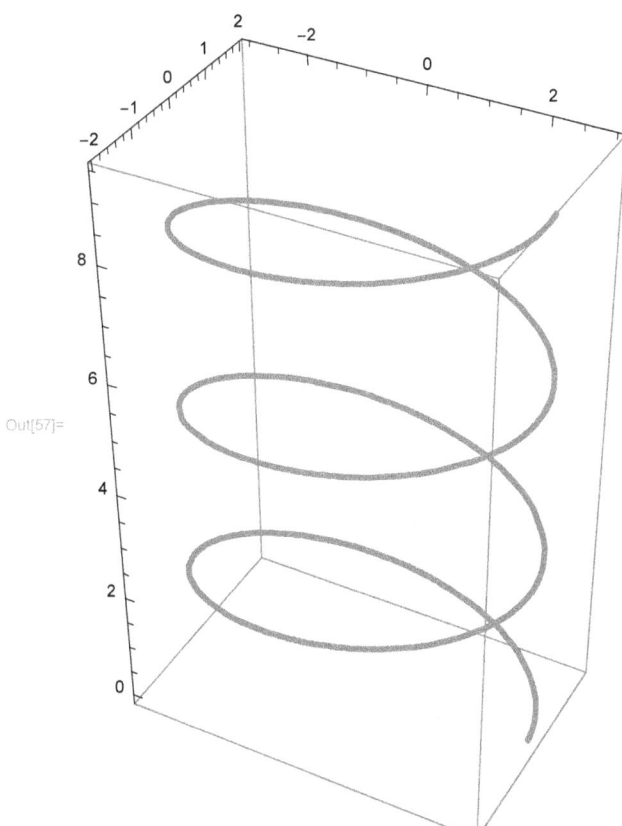

Out[57]=

1.5 Calculus

The **Limit** command can be used to find the limiting value of a function when the independent variable approaches a particular value. For instance, compute the following

$$\lim_{x \to 3} (x - 3) \csc(\pi x),$$

In[58]:= `Limit[(x - 3) Csc[π x], x → 3]`

Out[58]= $-\dfrac{1}{\pi}$

and

$$\lim_{x \to 0} (1 + \sin x)^{\cot(2x)}.$$

In[59]:= `Limit[(1 + Sin[x])^Cot[2 x], x → 0]`

Out[59]= $\sqrt{e}$

If **f[x]** represents a function, its derivative is represented by **f'[x]**. Higher derivatives are represented by **f''[x]**, **f'''[x]**, and so on.

In[60]:= `Clear[f]`

In[61]:= `f[x_] := Cosh[x^2 - 3 x + 1]`

In[62]:= `f'[x]`

Out[62]= $(-3 + 2 x) \, \text{Sinh}\left[1 - 3x + x^2\right]$

Another method to find the derivative of a function is by using the **D** command

In[63]:= `D[f[x], x]`

Out[63]= $(-3 + 2 x) \, \text{Sinh}\left[1 - 3x + x^2\right]$

The second derivative of $f(x) = \cosh\left(x^2 - 3x + 1\right)$

In[64]:= `f''[x]`

Out[64]= $(-3 + 2 x)^2 \, \text{Cosh}\left[1 - 3x + x^2\right] + 2 \, \text{Sinh}\left[1 - 3x + x^2\right]$

Or equivalently by using the **D** command

In[65]:= `D[f[x], {x, 2}]`

Out[65]= $(-3 + 2 x)^2 \, \text{Cosh}\left[1 - 3x + x^2\right] + 2 \, \text{Sinh}\left[1 - 3x + x^2\right]$

The third derivative of $f(x) = \cosh\left(x^2 - 3x + 1\right)$

In[66]:= `f'''[x]`

Out[66]= $6 (-3 + 2 x) \, \text{Cosh}\left[1 - 3x + x^2\right] + (-3 + 2 x)^3 \, \text{Sinh}\left[1 - 3x + x^2\right]$

In[67]:= `D[f[x], {x, 3}]`

Out[67]= $6 (-3 + 2 x) \, \text{Cosh}\left[1 - 3x + x^2\right] + (-3 + 2 x)^3 \, \text{Sinh}\left[1 - 3x + x^2\right]$

In Mathematica, **Integrate** command computes antiderivatives. For instance, compute the integral

$$\int (x+2) \sin(x^2 + 4x - 6)\, dx.$$

In[68]:= `Integrate[(x + 2) Sin[x^2 + 4 x - 6], x]`

Out[68]= $-\frac{1}{2} \cos[6] \cos\left[4x + x^2\right] - \frac{1}{2} \sin[6] \sin\left[4x + x^2\right]$

We may use the traditional symbol such as

In[69]:= $\int (x+2)\, \text{Sin}[x^2 + 4x - 6]\, dx$

Out[69]= $-\frac{1}{2} \cos[6] \cos\left[4x + x^2\right] - \frac{1}{2} \sin[6] \sin\left[4x + x^2\right]$

Notice that Mathematica omitted the constant of integration *C* from the answer.

The **Integrate** command can also be used to compute definite integrals

$$\int_0^\infty \frac{1}{1+x^2}\, dx.$$

In[70]:= `Integrate[1/(1 + x^2), {x, 0, ∞}]`

Out[70]= $\dfrac{\pi}{2}$

Or using the traditional symbol

In[71]:= $\int_0^\infty \dfrac{1}{1+x^2}\, dx$

Out[71]= $\dfrac{\pi}{2}$

Here is another example

In[72]:= $\int_0^{\frac{1}{\sqrt{2}}} \dfrac{x\, \text{ArcSin}[x^2]}{\sqrt{1 - x^4}}\, dx$

Out[72]= $\dfrac{\pi^2}{144}$

1.6 Differential Equations

```
In[73]:= (* First clear any values that may already have been
          assigned to the names of the various objects to be calculated *)
        Clear["Global`*"]
```

To solve differential equations for analytical solutions, Mathematica provides **DSolve** command to find the solutions.

Example 1. To solve the first-order differential equation $\frac{dy}{dx} - y^2 \cos x = 0$, we simply type

```
In[74]:= sol = DSolve[y'[x] - y[x]^2 Cos[x] == 0, y[x], x]
```

$$\text{Out[74]}= \left\{\left\{y[x] \to \frac{1}{-C[1] - \sin[x]}\right\}\right\}$$

Notice that the result is given as a list and we assign it to **sol**. The formula for the solution is in the form of replacement. Thus, to extract the solution we use **ReplaceAll** (/.) command.

```
In[75]:= y[x] /. sol
```

$$\text{Out[75]}= \left\{\frac{1}{-C[1] - \sin[x]}\right\}$$

Example 2. Find the solution of the second-order differential equation $2\frac{d^2x}{dt^2} + \frac{dx}{dt} + x = 0$, with initial conditions $x(0) = 1$, $x'(0) = 0$.

```
In[76]:= sol = DSolve[{2 x''[t] + x'[t] + x[t] == 0, x[0] == 1, x'[0] == 0}, x[t], t]
```

$$\text{Out[76]}= \left\{\left\{x[t] \to \frac{1}{7} e^{-t/4} \left(7 \cos\left[\frac{\sqrt{7}\, t}{4}\right] + \sqrt{7} \sin\left[\frac{\sqrt{7}\, t}{4}\right]\right)\right\}\right\}$$

Notice that we put the initial conditions together with the differential equation inside curly brackets. We can simplify the list by using **Flatten** command

```
In[77]:= sol = sol // Flatten
```

$$\text{Out[77]}= \left\{x[t] \to \frac{1}{7} e^{-t/4} \left(7 \cos\left[\frac{\sqrt{7}\, t}{4}\right] + \sqrt{7} \sin\left[\frac{\sqrt{7}\, t}{4}\right]\right)\right\}$$

To extract the solution we use **ReplaceAll** (**/.**) command

In[78]:= `x[t] /. sol`

Out[78]= $\frac{1}{7} e^{-t/4} \left(7 \cos\left[\frac{\sqrt{7}\, t}{4}\right] + \sqrt{7}\, \sin\left[\frac{\sqrt{7}\, t}{4}\right]\right)$

In[79]:= `(* Plot the analytical solution *)`
`Plot[x[t] /. sol, {t, 0, 6 π}, PlotRange → All, PlotStyle → Thickness[0.005]]`

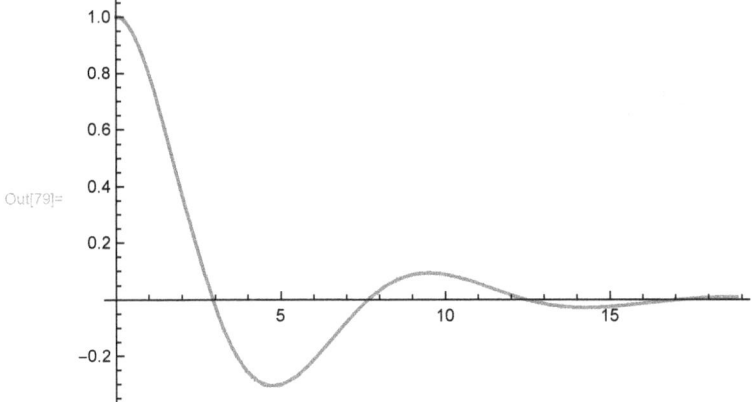

If the differential equation cannot be solve analytically, Mathematica provides the command **NDSolve** to obtain numerical solutions.

Example 3. Find the solution of the differential equation $\frac{dy}{dx} = 1 + y^2$ with initial condition $y(0) = 1$.

First we try with **DSolve** command.

In[80]:= `DSolve[{x''[t] + Cos[x[t]] == 0, x[0] == 0, x'[0] == 1}, x[t], t]`

--- Solve: Inverse functions are being used by Solve, so some solutions may not be found; use Reduce for complete solution information.

--- DSolve: For some branches of the general solution, unable to solve the conditions.

--- DSolve: For some branches of the general solution, unable to solve the conditions.

Out[80]= `{}`

However, Mathematica complains that DSolve is unable to solve the differential equation.
Let us now try using numerical approximation

In[81]:= **numsol = NDSolve[{x''[t] + Cos[x[t]] == 0, x[0] == 0, x'[0] == 1}, x[t], {t, 0, 10}]**

Out[81]= $\{\{x[t] \to \text{InterpolatingFunction}[\begin{smallmatrix}\text{Domain: }\{\{0., 10.\}\}\\ \text{Output: scalar}\end{smallmatrix}][t]\}\}$

Notice that to perform the integration, **NDSolve** command needs the initial conditions and the desired range of solution, in this case $0 \leq t \leq 10$.

In[82]:= **(* Plot the numerical solution *)**
Plot[x[t] /. numsol, {t, 0, 10}, PlotStyle → Thickness[0.005]]

Out[82]=

Chapter 2. Linear Oscillatory Systems

Initialization

```
In[1]:= Clear["Global`*"]
```

```
In[2]:= SetOptions[Plot, PlotStyle → AbsoluteThickness[2]];
       SetOptions[Plot3D, PlotStyle → AbsoluteThickness[2]];
       SetOptions[ParametricPlot, PlotStyle → AbsoluteThickness[2]];
       SetOptions[ParametricPlot3D, PlotStyle → AbsoluteThickness[2]];
```

2.1 Linear Simple Harmonic Motion

We will begin our analysis of the simplest linear oscillator system, a simple harmonic oscillator.

Consider a particle free to move along the x-axis. Let it be subject to a linear restoring force directed toward the origin O and proportional in magnitude to the displacement x of the particle. Then the force acting on the particle can be written as

$$F = -k x,$$

where k is the magnitude of the force acting on the particle when displaced a unit distance from the origin.

According to Newton's second law, the motion is described by a second-order, linear differential equation with constant coefficients,

(2.1.1) $$m \frac{d^2 x}{dt^2} = -k x$$

or

(2.1.2) $$\frac{d^2 x}{dt^2} = -\frac{k}{m} x \quad ,$$

where m is the mass of the particle.

```
In[6]:= (* Equation 2.1.2 *)
       equation = D[x[t], {t, 2}] == - k/m x[t]
```

$$Out[6]= \; x''[t] == -\frac{k\, x[t]}{m}$$

Chapter 2. Linear Oscillatory Systems

The essential characteristic of simple harmonic oscillator is that the **period** of the motion, that is, the time required for a particular configuration (both position and velocity) to repeat itself, is independent of the displacement from equilibrium. We shall choose the origin to be the point of equilibrium.

If we define $\omega_0^2 \equiv \dfrac{k}{m}$, the equation of motion becomes

In[7]:= **equation1 = $\left(\text{equation} \;/.\; \dfrac{k}{m} \to \omega 0^2\right)$**

Out[7]= $x''[t] == -\omega 0^2\, x[t]$

where ω_0 is called the angular frequency, or the number of oscillations in 2π seconds.

The value of ω_0 does not depend on the initial conditions but determined by the parameters of the oscillatory system, where in this case k and m.

The solution to this differential equation can be found by using **DSolve** command

In[8]:= **solution1 = DSolve[equation1, x, t]**

Out[8]= $\{\{x \to \text{Function}[\{t\},\, C[1]\, \text{Cos}[t\, \omega 0] + C[2]\, \text{Sin}[t\, \omega 0]]\}\}$

where C[1] and C[2] are constants of integration. These constants are determined by the initial conditions of the system.

Consider $\omega_0 = 2$, C[1] = 1, C[2] = 1, let us plot the solution of the equation of motion

In[9]:= **Plot[x[t] /. solution1 /. {$\omega 0 \to 2$, C[1] $\to 1$, C[2] $\to 1$},
{t, 0, 4 π}, AxesLabel $\to$ {"t", "x[t]"}, BaseStyle $\to$ {FontSize $\to$ 12}]**

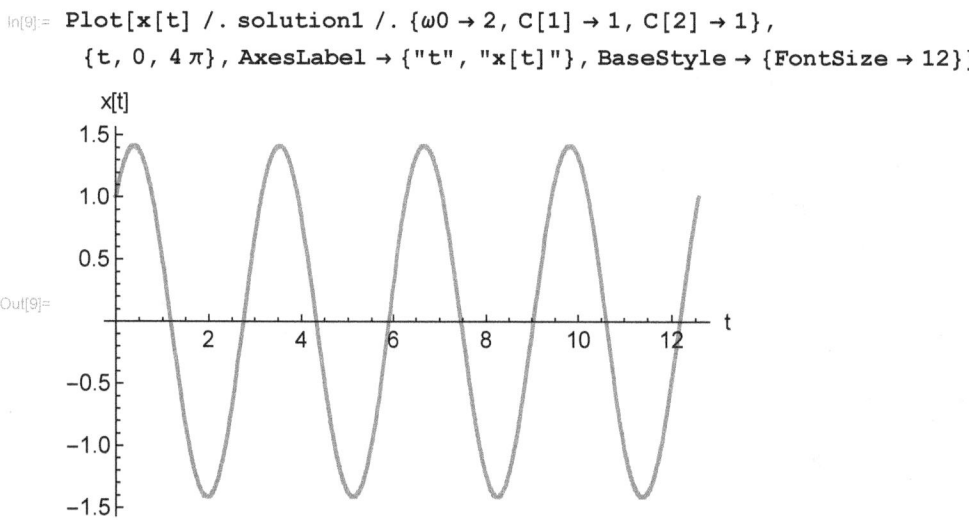

Figure 2.1.1 The graph of the solution of $\frac{d^2x}{dt^2} = -\frac{k}{m} x$, in which $\omega_0 = 2$, $C[1] = 1$, $C[2] = 1$

The solution of Eq. 2.1.2 is a sinusoidal oscillation about the equilibrium $x = 0$, as is shown in the Figure 2.1.1.

The motion exhibits the following features:

(1). It is characterized by an angular frequency ω_0.

The motion repeats itself after the angular argument of the sinusoidal functions ($\omega_0 t$) advances by 2π. The time required of phase advance of 2π is $T_0 = \frac{2\pi}{\omega_0}$. We can confirm this using Mathematica

In[10]:= `period = Solve[ω0 (t + T0) == ω0 t + 2 π, T0] // Flatten`

Out[10]= $\left\{ T0 \to \frac{2\pi}{\omega 0} \right\}$

Thus,

In[11]:= `T0 = T0 /. period`

Out[11]= $\frac{2\pi}{\omega 0}$

T_0 is called the **period** of the motion. We notice here that it does not depend on either C[1] or C[2], so it is independent of initial conditions.

The frequency f_0 is given by $\frac{1}{T_0}$, so $f_0 = \frac{\omega_0}{2\pi}$.

In[12]:= `f0 = `$\frac{1}{T0}$

Out[12]= $\frac{\omega 0}{2\pi}$

(2). The motion is bounded. It is confined within the limits $-A \le x \le A$.
A is the maximum displacement from the equilibrium. It is called the **amplitude** of the motion.

(3). The constants of integration C[1] and C[2] determines the initial displacement x at time $t = 0$. For example at $t = 0$, we have

Chapter 2. Linear Oscillatory Systems

In[13]:= `x[t] /. solution1 /. t → 0`

Out[13]= `{C[1]}`

the initial displacement, $x(0) = C[1]$.

An oscillatory motion does not occur, when $x(0) = 0$ and $\dot{x}(0) = 0$, for then the oscillator is in equilibrium state and remains so.

The maximum displacement from equilibrium are determined by initial conditions can be computed using **Maximize** command

In[14]:= `max = Maximize[x[t] /. solution1 /. {ω0 → 2, C[1] → 1, C[2] → 1}, t] // N`

Out[14]= `{1.41421, {t → 0.392699}}`

The maximum displacement A is 1.41421, first occurs at $t_{max} = 0.392699$.
Since the motion repeats itself by T_0, the successive maximum displacement occurs at
$$t_n = t_{max} + n\, T_0,$$
where $n = 0,1,2,3,\ldots$

In[15]:= `(* Compute the maximum point coordinates for n=0,1,2,3 *)`
`maxpoints = ({(t /. max[[2]]) + #1 * (T0 /. ω0 → 2), max[[1]]} &) /@ {0, 1, 2, 3}`

Out[15]= `{{0.392699, 1.41421}, {3.53429, 1.41421}, {6.67588, 1.41421}, {9.81748, 1.41421}}`

In[16]:= `(* Plot the maximum point coordinates on the graph *)`
`Plot[x[t] /. solution1 /. {ω0 → 2, C[1] → 1, C[2] → 1},`
` {t, 0, 4 π}, Epilog → {PointSize[0.02], Red, Point[maxpoints]},`
` AxesLabel → {"t", "x[t]"}, BaseStyle → {FontSize → 12}]`

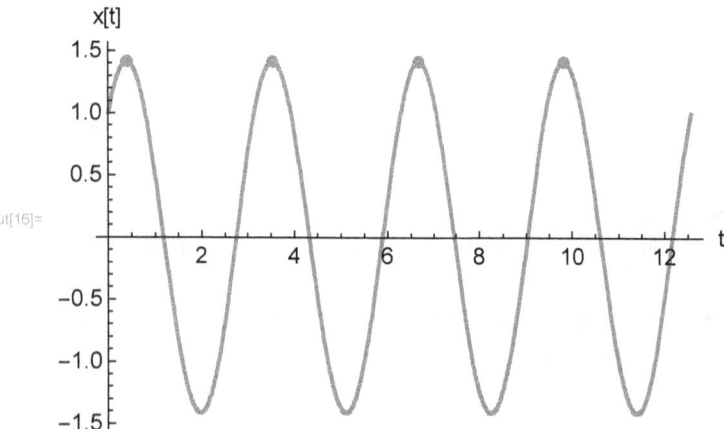

Figure 2.1.2 The graph of the solution of $\frac{d^2x}{dt^2} = -\frac{k}{m}x$, with maximum points, where $\omega_0 = 2$, $C[1] = 1$, $C[2] = 1$

2.1.1 Energy Considerations in Simple Harmonic Motion

Let us calculate the work done by the linear restoring force in moving the particle from the equilibrium position ($x = 0$) to some position x, which equals to the potential energy

$$V(x) = \int_0^{x(t)} k\, x\, dx$$

In[17]:= `V = `$\int_0^{x[t]} k\, x\, dx$

Out[17]= $\frac{1}{2} k\, x[t]^2$

Notice that the potential energy is quadratic in position variable.

In[18]:= `(* Potential energy of the particle *)`
`V1 = V /. solution1 // Simplify`

Out[18]= $\left\{\frac{1}{2} k\, (C[1] \cos[t\, \omega 0] + C[2] \sin[t\, \omega 0])^2\right\}$

The kinetic energy is quadratic in velocity variable, thus

In[19]:= `(* Kinetic energy of the particle *)`
`K = `$\frac{1}{2}$` m (D[x[t], t])`^{2}

Out[19]= $\frac{1}{2} m\, x'[t]^2$

In[20]:= `(* Kinetic energy of the particle`
   ` Applying it to our solution of x'[t], we obtain`
   ` the kinetic energy`
`*)`
`K1 = K /. solution1`

Out[20]= $\left\{\frac{1}{2} m\, (\omega 0\, C[2] \cos[t\, \omega 0] - \omega 0\, C[1] \sin[t\, \omega 0])^2\right\}$

The total energy is given by the sum of kinetic and potential energies

In[21]:= `(* Total energy of the particle *)`
`totalEnergy = E1 == (K1 + V1) /. (k → ω0² m) // Simplify`

Out[21]= $E1 == \left\{\frac{1}{2} m\, \omega0^2 \left(C[1]^2 + C[2]^2\right)\right\}$

The total energy depends on initial conditions of the oscillator, that is, it is proportional to the square of C[1] and C[2], which constitute the amplitude of *x*.

Notice also that the total energy is constant if there is no other forces except the restoring force acting on the particle. Since the total energy is independent of time, it is **conserved**.

The relation between total energy and initial conditions can be expressed as follows

In[22]:= `initialConditions = Solve[totalEnergy, C[1]]`

Out[22]= $\left\{\left\{C[1] \to -\frac{i\sqrt{-2\,E1 + m\,\omega0^2\,C[2]^2}}{\sqrt{m}\,\omega0}\right\}, \left\{C[1] \to \frac{i\sqrt{-2\,E1 + m\,\omega0^2\,C[2]^2}}{\sqrt{m}\,\omega0}\right\}\right\}$

Example. Calculate the *average* kinetic, potential, and total energies of the simple harmonic oscillator and show that it is equal to the instantaneous total energy obtained above.

Solution

Let T_0 be the period of the harmonic oscillator. We define the average kinetic energy to be

In[23]:= `SetAttributes[T0, Constant]`

$\langle K \rangle = \frac{1}{T0} \int_0^{T0} K1\, dt$

Out[24]= $\left\{\frac{1}{4} m\, \omega0^2 \left(C[1]^2 + C[2]^2\right)\right\}$

Notice that we have set the attribute of T_0 to constant using **SetAttributes** command.

Similarly, we compute the average potential energy as

In[25]:= $\langle V \rangle = \frac{1}{T0} \int_0^{T0} V1 \, dt$

Out[25]= $\left\{\frac{1}{4} k \left(C[1]^2 + C[2]^2\right)\right\}$

But since $\omega_0^2 \equiv \frac{k}{m}$, we replace k with $m\omega_0^2$

In[26]:= $\langle V \rangle = \langle V \rangle \, /. \, k \to m\,\omega 0^2$

Out[26]= $\left\{\frac{1}{4} m\,\omega 0^2 \left(C[1]^2 + C[2]^2\right)\right\}$

Here we notice that the average kinetic energy and the potential energy are equal.

So, the average total energy is

In[27]:= $\langle E \rangle = \langle K \rangle + \langle V \rangle$

Out[27]= $\left\{\frac{1}{2} m\,\omega 0^2 \left(C[1]^2 + C[2]^2\right)\right\}$

Therefore, the average total energy of the oscillator is equal to its total instantaneous energy.

2.1.2 Representation on the Phase Plane

We shall study the motion of a harmonic oscillator by representing this motion on a two-dimensional plane, where x is the horizontal coordinates and $\dot{x}$ is the vertical coordinates. To each state of our system, that is, to each pair of values of the coordinate x and velocity $\dot{x}$, there corresponds a point on the x-$\dot{x}$ plane. This plane is called a **phase plane** or a two-dimensional phase space. To each point $(x, \dot{x})$ on the phase plane there corresponds one and only one state of the system.

As time varies, the point $(x, \dot{x})$, which describes the state of the oscillating particle, will trace out certain phase path in the phase plane. For different initial conditions, the motion will be described by different **phase paths**. Any given path corresponds to a certain set of initial conditions, represents the time history of the oscillating particle. It must not be confused with the actual trajectory of motion. All possible phase paths constitutes the **phase diagram** of the oscillator. The velocity of such representative point is called the **phase velocity**; again this must not be confused with the actual velocity.

First, we define the coordinates $(x, \dot{x})$ of the phase plane, in which $\dot{x}$ is the time derivative of the displacement *x*. The nested list is converted to a list of ordered pairs with **Flatten** command. Then we assign the coordinates to **phasePoints**.

```
In[28]:= (* Define the phase points *)
        phasePoints = Flatten[{x[t], D[x[t], t]} /. solution1]
Out[28]= {C[1] Cos[t ω0] + C[2] Sin[t ω0], ω0 C[2] Cos[t ω0] - ω0 C[1] Sin[t ω0]}
```

Using **Table** command, we store the coordinates corresponding to various values of C[1] in the range of 1 to 5, into a table and draw the graphs using **ParametricPlot** command. Instead of displaying the results directly we assign them to **phaseDiagramPlots**. Finally, using **TabView** command we display the graphs in tabs. Each tab shows different paths at different time ranging from 0.5 to 4.2π. The final result is shown in **Figure 2.1.3**. Click on any tab to view different paths corresponding to different values of C[1] at that particular time.

```
In[29]:= (* Plot for various initial conditions,
         a phase diagram by continuously changing the time.
             Consider ω₀=1/2, C[2]=1.
         *)
         phaseDiagramplots =
           (ParametricPlot[Evaluate[Table[phasePoints /. {C[1] → i, C[2] → 1, ω0 → 1/2},
                 {i, 1, 5}]], {t, 0, #1}, PlotRange → ({-7, 7}, {-4, 4}),
              AxesLabel → {"x", "ẋ"}] &) /@ Table[tvar, {tvar, 0.5, 4.2 π, 0.5}];
         TabView[Table[k → phaseDiagramplots[[k]], {k, Length[phaseDiagramplots]}]]
```

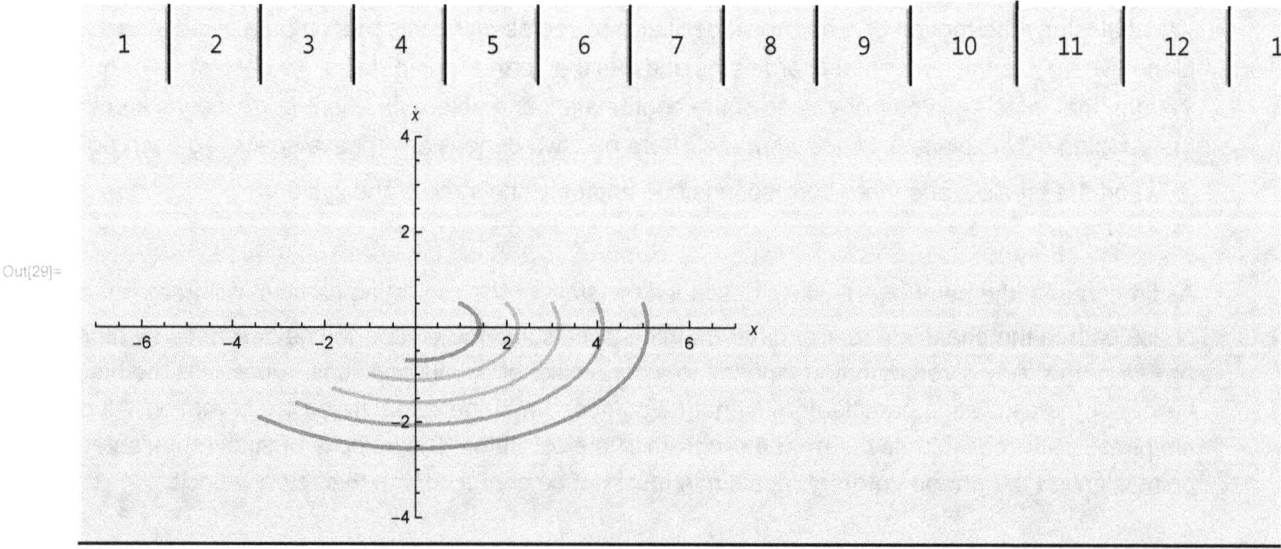

Figure 2.1.3 The phase paths for various initial conditions. Each tab show the paths for different time

($\omega_0 = 1/2$, $C[2] = 1$)

Let us now draw the complete phase diagrams for various initial conditions

In[30]:=
```
(* Plot for various initial conditions,
a complete phase diagram. Set ω₀=1/2, C[2]=1.
*)
ParametricPlot[Evaluate[Table[phasePoints /. {C[1] → i, C[2] → 1, ω0 → 1/2}, {i, 1, 5}]],
  {t, 0, 4.2 π}, PlotRange → (-7  7
                              -4  4),
  AxesLabel → {"x", "ẋ"}, BaseStyle → {FontSize → 14}]
```

Out[30]=

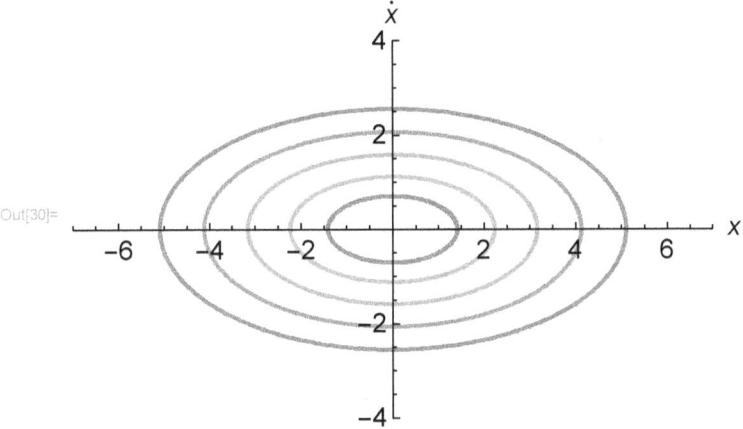

Figure 2.1.4 The phase diagrams for various initial conditions ($\omega_0 = 1/2$, $C[2] = 1$)

As can be seen from **Figure 2.1.4**, the motion of the representative point will always be in a **clockwise** direction, because for $x > 0$ the velocity x' is always decreasing and for $x < 0$ the velocity x' is always increasing.

A time-evolved phase path can be drawn by using **ParametricPlot3D** command. Notice that we make the phase coordinates three-dimensional by appending t to **phasePoints**, using the **Join** command.

In[31]:=
```
ParametricPlot3D[Evaluate[Join[phasePoints /. {C[1] → 5, C[2] → 1, ω0 → 1/2}, {t}]],
  {t, 0, 30}, PlotRange → All, PlotStyle → {Blue, AbsoluteThickness[2]},
  PlotLabel → "Time-evolved Phase Path", AxesLabel → {"x", "ẋ", "t"},
  BoxRatios → {1, 1, 1}, BaseStyle → {FontSize → 12}]
```

Out[31]=
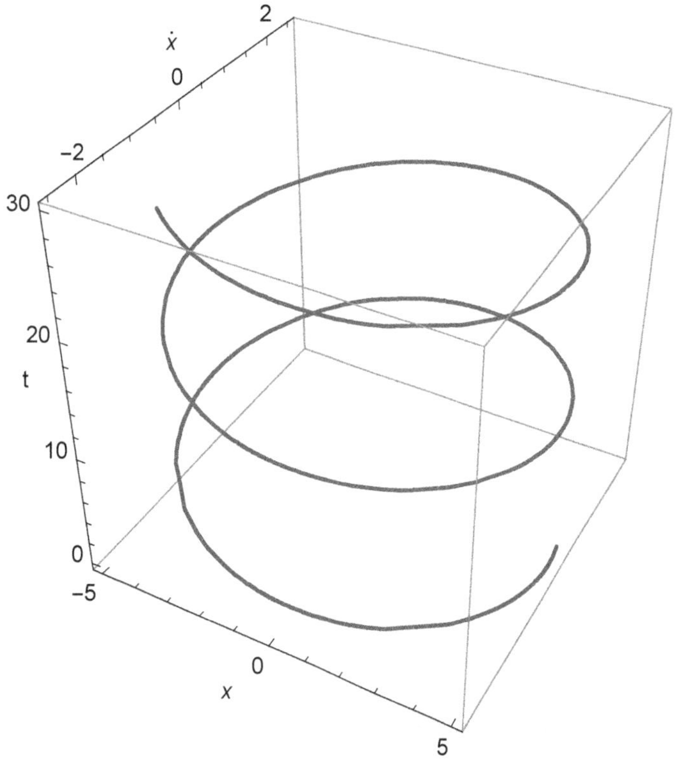

Figure 2.1.5 The time-evolved phase path in three-dimensional space

Knowing the solution of the differential equation of the harmonic oscillator, we can easily find the equation of the path on the phase plane. As a matter of fact, the equations of the phase points

In[32]:=
```
phasePoints
```

Out[32]=
```
{C[1] Cos[t ω0] + C[2] Sin[t ω0], ω0 C[2] Cos[t ω0] - ω0 C[1] Sin[t ω0]}
```

are the parametric equations of the phase path. By eliminating the parameter t from these equations, we can compute the coordinate equations of a path.

In[33]:= **phaseCurve = Assuming[{ω0² ≠ 0 && C[1] > 0 && C[2] > 0},**
FullSimplify[Eliminate[Thread[{x, ẋ} == phasePoints], t]]]

— Eliminate: Inverse functions are being used by Eliminate, so some solutions may not be found; use Reduce for complete solution information.

Out[33]= $x^2 \, \omega 0^2 + \dot{x}^2 == \omega 0^2 \left(C[1]^2 + C[2]^2 \right)$

It is easily seen that as C[1] and C[2] varies, this is the equation of a family of similar **ellipses**. All these ellipses represent paths of motion of the phase point.

Thus, the equation of a phase path can be written as

In[34]:= **phaseCurve = phaseCurve /. initialConditions // Simplify // Union**

Out[34]= $\left\{ \dfrac{2 \, E1}{m} == x^2 \, \omega 0^2 + \dot{x}^2 \right\}$

This is the equation of an ellipse for a given total energy **E1**, for the given initial conditions C[1] and C[2].

In[35]:= **(* Solve for the total energy *)**
phaseCurve1 = Solve[phaseCurve, E1]

Out[35]= $\left\{ \left\{ E1 \to \dfrac{1}{2} \, m \left(x^2 \, \omega 0^2 + \dot{x}^2 \right) \right\} \right\}$

We have shown in **Section 2.1.1** that the total energy of a simple harmonic oscillator is conserved. Let us now represents the conservation of total energy in the phase plane.

```mathematica
In[36]:= ContourPlot[Evaluate[E1 /. phaseCurve1 /. {m → 1, ω0 → 1/2}],
   {x, -5, 5}, {ẋ, -3, 3}, Axes → True, AxesLabel → {"x", "ẋ"},
   PlotLabel → "Energy Representations in the Phase Plane"]
```

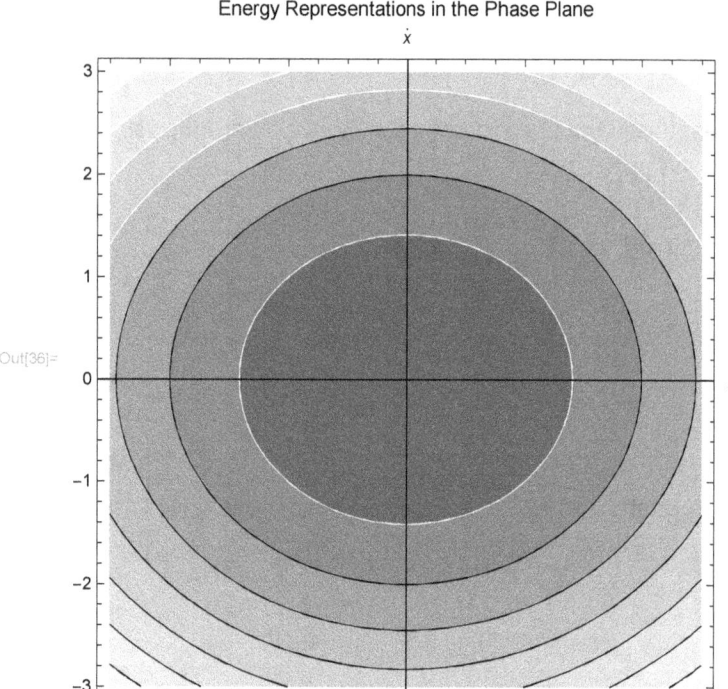

Figure 2.1.6 Level curves of the total energy

As can be seen from the figure above the phase plane is filled with nested ellipses. Each ellipse of phase path represents a particular value of the conserved total energy. The point at the center $\{x(t) = 0,\ x'(t) = 0\}$ represents the position of equilibrium and corresponds to zero energy.

Let us find the value of the phase velocity. We shall introduce the phase position vector $\mathbf{r}_p \equiv x\,\hat{\mathbf{i}} + \dot{x}\,\hat{\mathbf{j}}$.

```mathematica
In[37]:= r_p = phasePoints
Out[37]= {C[1] Cos[t ω0] + C[2] Sin[t ω0], ω0 C[2] Cos[t ω0] - ω0 C[1] Sin[t ω0]}
```

The phase velocity is defined as

```mathematica
In[38]:= v_p = D[r_p, t]
Out[38]= {ω0 C[2] Cos[t ω0] - ω0 C[1] Sin[t ω0], -ω0² C[1] Cos[t ω0] - ω0² C[2] Sin[t ω0]}
```

Since sine and cosine functions never reduces to zero at the same time, the phase velocity never reduces to zero unless both C[1] and C[2] are zero.

2.1.3 Two-dimensional Harmonic Oscillator

In section 2.1.1 we study one-dimensional motion of a particle moving along the x-axis, subject to a linear restoring force directed toward the origin O and proportional in magnitude to the displacement x of the particle. We can extend this motion to particle moving on the x-y plane such that the force becomes **F** = −k **r**, where k is the magnitude of the force acting on the particle when displaced a unit distance from the origin. Accordingly, the equation of motion becomes

(2.1.3.1) $$m \frac{d^2 \mathbf{r}}{dt^2} = -k\, \mathbf{r},$$

where $\mathbf{r} = x\,\hat{\mathbf{i}} + y\,\hat{\mathbf{j}}$.

Equation 2.1.3.1 is a differential equation for **linear isotropic oscillator** since the restoring force is independent of the direction of the displacement. The two components of the differential equations can then be written separately as

```
In[39]:= xIsotropic = D[x[t], {t, 2}] == - k/m x[t]
        yIsotropic = D[y[t], {t, 2}] == - k/m y[t]
```

Out[39]= $x''[t] == -\dfrac{k\, x[t]}{m}$

Out[40]= $y''[t] == -\dfrac{k\, y[t]}{m}$

Our analysis in previous sections for one-dimensional oscillator is applicable in this case for each component of the differential equations.

In general, however, the magnitudes of the components of the restoring force depend on the direction of the displacement. In such cases, we have **non-isotropic oscillators**. The differential equations for non-isotropic oscillators become

```
In[41]:= xNonisotropic = D[x[t], {t, 2}] == - k1/m x[t]
        yNonisotropic = D[y[t], {t, 2}] == - k2/m y[t]
```

Out[41]= $x''[t] == -\dfrac{k1\, x[t]}{m}$

Out[42]= $y''[t] == -\dfrac{k2\, y[t]}{m}$

The solutions for these separated equations can readily be written as

```
In[43]:= x[t_] := A1 Cos[ω1 t + ϕ1]
        y[t_] := A2 Cos[ω2 t + ϕ2]
```

in which $\omega_1 = \sqrt{\dfrac{k_1}{m}}$ is the angular frequency of x-component oscillator and $\omega_2 = \sqrt{\dfrac{k_2}{m}}$ is the angular frequency of the y-component oscillator. ϕ_1 and ϕ_2 are called the phases of the oscillators.

The constants A_1, A_2, ϕ_1, ϕ_2 are determined from the initial conditions of any given case.

We can show using Mathematica that the proposed solutions are indeed the solutions for the non-isotropic differential equations.

```
In[45]:= (* x[t] is indeed the solution for the
         nonisotropic x component differential equation *)
         xNonisotropic /. k1/m → ω1²
Out[45]= True
```

```
In[46]:= (* y[t] is indeed the solution for the
         nonisotropic y component differential equation *)
         yNonisotropic /. k2/m → ω2²
Out[46]= True
```

We notice that x(t) and y(t) are parametric equations in time t, thus, we can plot them using **ParametricPlot** command as follows

```
In[47]:= ParametricPlot[Evaluate[{x[t], y[t]} /. {A1 → 2, ω1 → 1, ϕ1 → 0, A2 → 1, ω2 → 1, ϕ2 → π/2}],
         {t, 0, 20 π}, PlotRange → {{-3, 3}, {-2, 2}}, AxesLabel → {"x", "y"}]
```

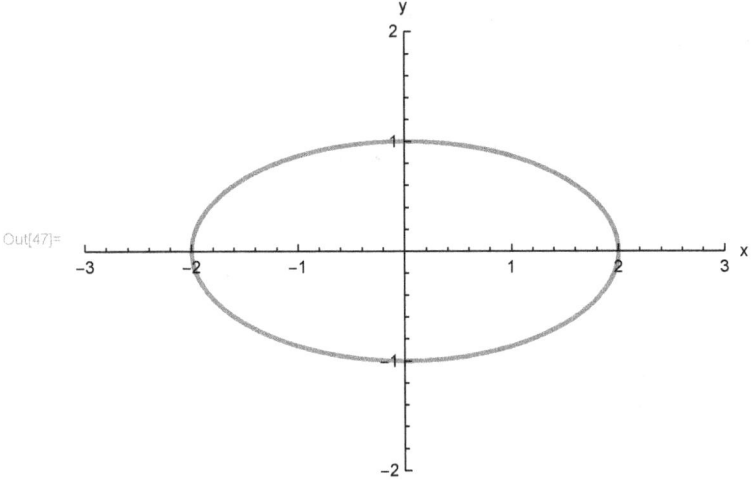

Out[47]=

Figure 2.1.7 The parametric curve of solutions, x(t) vs y(t) of the non-isotropic oscillators

Depending on the constants of integration or initial conditions, our parametric curves can be a line, circle, ellipse, parabola, or hyperbola. These family of curves are called **Lissajous curves**.

Consider $A_1 = A_2 = 1$, $\omega_1 = \omega_2 = 1$, $\phi_1 = 0$. Varying the phase ϕ_2 in the range from 0 to $\pi/2$ with increment 0.5, we use ParametricPlot command to draw several Lissajous curves, and assign them to **lissajousPlots**. Then using **TabView** command we represent the graphs on tabs.

```
In[48]:= lissajousPlots = (ParametricPlot[
         Evaluate[{x[t], y[t]} /. {A1 → 1, ω1 → 1, ϕ1 → 0, A2 → 1, ω2 → 1, ϕ2 → #1}],
         {t, 0, 20 π}, ImageSize → 200, PlotRange → Full] &) /@ Table[i, {i, 0, π/2, 0.5}];
         TabView[Table[k → lissajousPlots[[k]], {k, Length[lissajousPlots]}]]
```

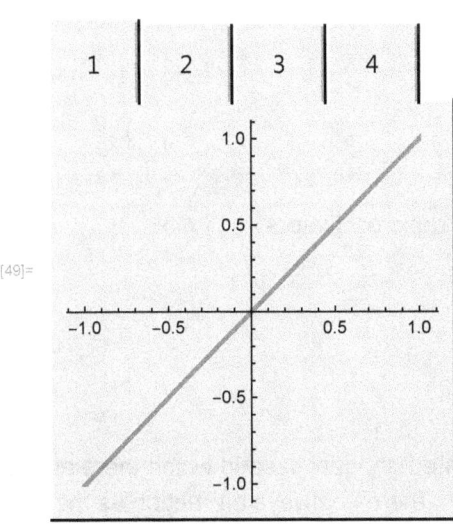

Out[49]=

Figure 2.1.8 The Lissajous curves of the non-isotropic oscillators.

In the following codes, we provide some interactive manipulations of the Lissajous curves. The reader can click on the plus signs at the right of the sliders to bring up the menus and manipulate the interactions to see how the parameters determine the Lissajous curves.

```
In[50]:= Manipulate[ParametricPlot[Evaluate[{A1 Cos[ω1 t + ϕ1], A2 Cos[ω2 t + ϕ2]}],
    {t, 0, 2 π}, PlotRange → {-3, 3}, PerformanceGoal → "Quality",
    Frame → True, FrameStyle → Directive[Red]],
   {{ω1, 1}, 1, 10, 1, Appearance → "Labeled"}, {{A1, 1}, 1, 3, Appearance → "Labeled"},
   {{ϕ1, 0}, 0, 2 π, Appearance → "Labeled"},
   {{ω2, 1}, 1, 10, 1, Appearance → "Labeled"}, {{A2, 1}, 0, 3, Appearance → "Labeled"},
   {{ϕ2, π/2}, 0, 2 π, Appearance → "Labeled"}, ControlPlacement → Right]
```

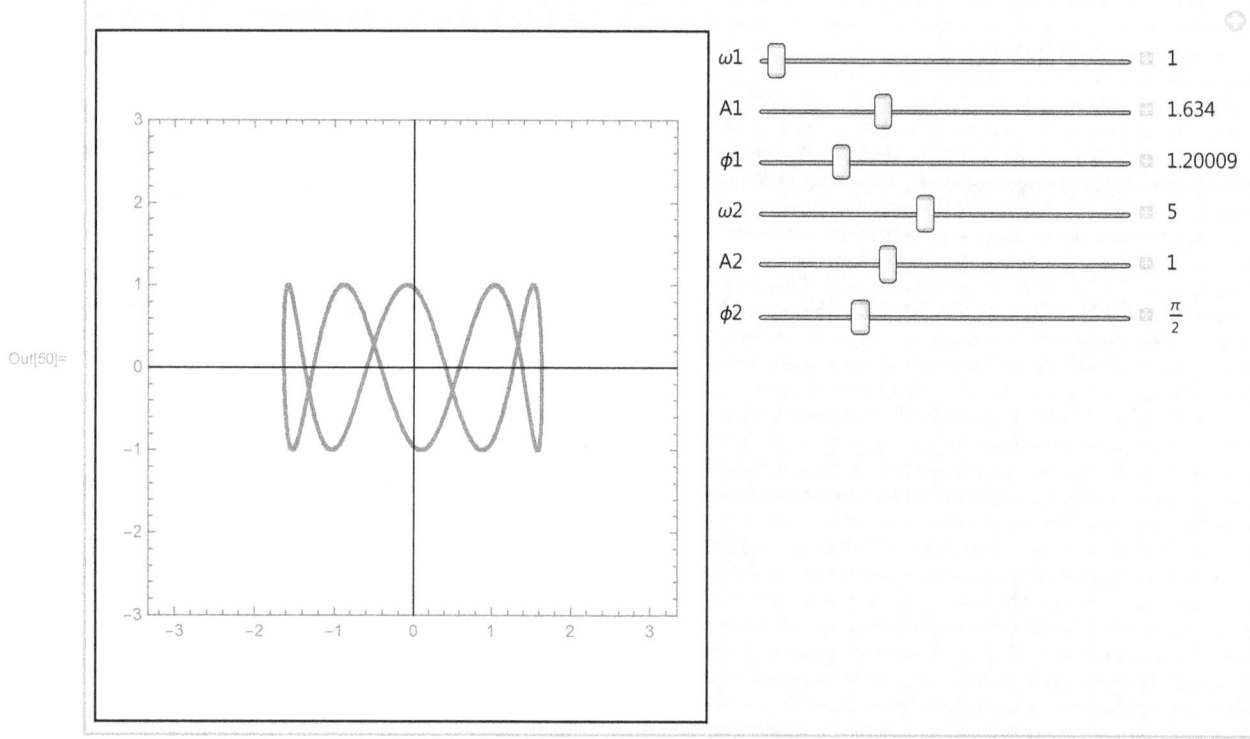

Figure 2.1.9 The animations of Lissajous curves of the nonisotropic oscillators.

2.2 Linear Oscillator in the Presence of Frictional Force

In our previous analysis of harmonic oscillator, we have presented an ideal case in which there are no frictional forces acting on the system. However, in real situation, there is always air friction as the oscillator moves through the air. The motion is damped as all real oscillating systems are subject to damping forces and eventually will cease to oscillate, and such damped oscillations are the subject of this section.

We shall assume that the frictional force is proportional to velocity and for small velocity, this assumption is found in satisfactory agreement with experiment.

For our particle as described in Section 2.1, the equation of motion becomes

(2.2.1) $$m \frac{d^2x}{dt^2} = -kx - \gamma \dot{x},$$

in which the damping force is represented by the term $-\gamma \dot{x}$, where γ is a constant called the **damping factor** which measures the strength of the damping force.

```
In[51]:= Clear["Global`*"]
```

```
In[52]:= (* Eq. 2.2.1 *)
        equation = m D[x[t], {t, 2}] == -k x[t] - γ D[x[t], t]
```
Out[52]= $m\, x''[t] == -k\, x[t] - \gamma\, x'[t]$

The equation of motion in the presence of frictional force can be re-written as,

(2.2.2) $$\frac{d^2x}{dt^2} + \frac{\gamma}{m} \dot{x} + \frac{k}{m} x = 0$$

```
In[53]:= (* Eq. 2.2.2 *)
        equation = D[x[t], {t, 2}] + 2 β D[x[t], t] + ω0^2 x[t] == 0
```
Out[53]= $\omega 0^2\, x[t] + 2\, \beta\, x'[t] + x''[t] == 0$

Here, we define $\beta \equiv \frac{\gamma}{2m}$, as the damping factor, $\omega_0 \equiv \sqrt{\frac{k}{m}}$ is the **natural frequency** in the absence of damping.

The parameters β, γ, and ω_0 are positive real constants. The solution of this equation can be obtained using **DSolve** command

```
In[54]:= solution = Flatten[DSolve[equation, x, t]]
```
Out[54]= $\left\{ x \to \text{Function}\left[\{t\}, e^{t\left(-\beta - \sqrt{\beta^2 - \omega 0^2}\right)} C[1] + e^{t\left(-\beta + \sqrt{\beta^2 - \omega 0^2}\right)} C[2]\right]\right\}$

Notice the term $\sqrt{\beta^2 - \omega_0^2}$ that appears in the solution.

Thus, there are three possible cases:

Case 1. Underdamping: $\quad \omega_0^2 > \beta^2$

Case 2. Critical damping: $\quad \omega_0^2 = \beta^2$

Case 3. Overdamping: $\quad \omega_0^2 < \beta^2$

2.2.1 Case 1. Underdamped Motion

In this case, β is small enough so that $\omega_0^2 > \beta^2$.

The exponents of the solution becomes imaginary. A mass initially displaced and then released from rest oscillates while with the presence of a real factor β in the exponent of the solution will eventually lead to the gradual death of the oscillatory motion.

If we define $\omega_1^2 \equiv \omega_0^2 - \beta^2$, where $\omega_1^2 > 0$. The solution reduces to

In[55]:= `underdampedSolution = PowerExpand[x[t] /. solution /. ω0^2 → β^2 + ω1^2]`

Out[55]= $e^{t(-\beta - i\,\omega 1)} C[1] + e^{t(-\beta + i\,\omega 1)} C[2]$

where ω_0 and ω_1 are the angular frequencies of the **undamped** and **underdamped** harmonic oscillators, respectively.

The solution looks similar to the undamped oscillator. However, in our present case, the amplitude decreases with time because of the factor $e^{-\beta t}$, which forms a curve called the **envelope** of the displacement versus time.

In[56]:= `envelope = underdampedSolution /. ω1 → 0`

Out[56]= $e^{-t\,\beta} C[1] + e^{-t\,\beta} C[2]$

Let us now plot the solution with the envelope of the displacement versus time

```
In[57]:= Plot[Evaluate[{envelope, -envelope, underdampedSolution} /.
    {ω1 → 1, β → 1/7, C[1] → 1, C[2] → 1}], {t, -1, 15},
   AxesLabel → {"t", "x"}, PlotStyle → {{Blue, Dashed}, {Blue, Dashed}, {Red}},
   PlotRange → All, Epilog → {Text[Envelope, {5, 1.5}]}]
```

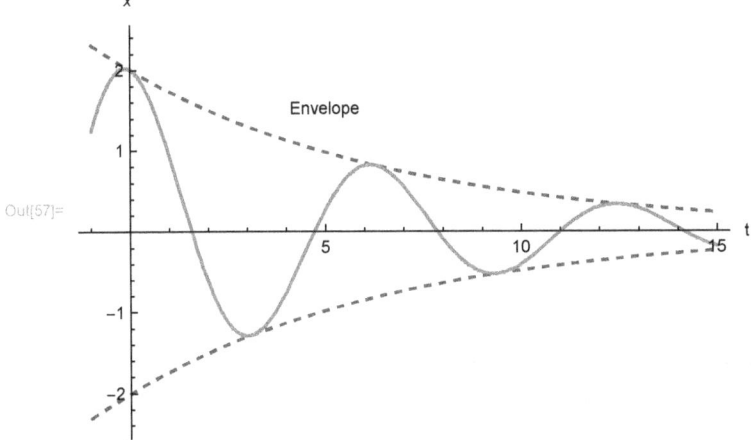

Figure 2.2.1 The graph of the solution with the envelope of the displacement versus time ($\omega_1 = 1$, $\beta = 1/7$, $C[1] = 1$, $C[2] = 1$)

The two curves $e^{-\beta t}$ and $-e^{-\beta t}$ form an envelope of the curve of motion while the oscillation takes place. The oscillator eventually returns to a state of rest at the equilibrium position due to the presence of damping force.

The solution, $x(t)$ is not a periodic function because of the presence of a factor $e^{-\beta t}$. Therefore we cannot speak of a period in the strict sense of this word. However, because the interval of time between the occurrence of two successive maxima remains constant, we define the period of the underdamped oscillator, T_d by

$$T_d \equiv \frac{2\pi}{w_1}.$$

```
In[58]:= periodUnderdamped = Td == (2π)/ω1 /. ω1 → Sqrt[ω0^2 - β^2]
```

$$\text{Out[58]= } T_d == \frac{2\pi}{\sqrt{-\beta^2 + \omega 0^2}}$$

Thus,

$$T_d = \frac{2\pi}{\sqrt{\omega_0^2 - \beta^2}}.$$

Note that the underdamped oscillator vibrates slower than the undamped oscillator because of the presence of the damping factor β.

For $\omega_1 = 1$, the amount of underdamped period becomes

```
In[59]:= periodUnderdamped = Td == (2 π)/ω1 /. ω1 → 1 // N
Out[59]= Td == 6.28319
```

The ratio of the amplitudes of the oscillation at two successive maxima is

```
In[60]:= (envelope /. t → t + Td)/envelope // Simplify
Out[60]= e^(-Td β)
```

Thus, in one complete period the amplitude diminishes by a factor of $e^{-\beta T_d}$, where βT_d is called the **logarithmic decrement**. Also, in time $t = \beta^{-1}$, the amplitude decays by a factor e^{-1}.

The amplitudes at various times can be calculated using **FindMaximum** command for our given conditions ($\omega_1 = 1$, $\beta = 1/7$, $C[1] = 1$, $C[2] = 1$). We then use **Table** command to list the amplitude vs time.

```
In[61]:= amplitude = (FindMaximum[underdampedSolution /.
           {ω1 → 1, β → 1/7, C[1] → 1, C[2] → 1}, {t, #1}] &) /@ {0, 6, 12};
amplitudedata = Table[{amplitude[[n, 1]], t} /. amplitude[[n, 2]], {n, 1, 3}];
amplitudedata = Prepend[amplitudedata, {"amplitude", "t"}];
Text@Grid[amplitudedata, Alignment → Left, Dividers → {Center, 2 → True}]
```

Out[64]=

amplitude	t
2.02044	-0.141897
0.823426	6.14129
0.335585	12.4245

Our data confirm that though the amplitude is diminishing, the period is constant. For our chosen initial conditions, it equals to 6.28319.

```
In[65]:= (t /. amplitude[[2, 2]]) - (t /. amplitude[[1, 2]])
Out[65]= 6.28319
```

2.2.1.1 Energy Considerations in Underdamped Motion

Just as the case in undamped harmonic motion in Section 2.1.1, the total energy of the damped oscillator (H) is given by the sum of kinetic energy (K) and potential energy (V) as

$$H = K + V = \frac{1}{2} m \dot{x}^2 + \frac{1}{2} k x^2$$

In[66]:= (* Kinetic Energy *)
$$K = \frac{1}{2} m \, (D[x[t], t])^2$$

Out[66]= $\frac{1}{2} m \, x'[t]^2$

In[67]:= (* Potential Energy *)
$$V = \frac{1}{2} k \, x[t]^2$$

Out[67]= $\frac{1}{2} k \, x[t]^2$

In[68]:= (* Total Energy *)
H = K + V

Out[68]= $\frac{1}{2} k \, x[t]^2 + \frac{1}{2} m \, x'[t]^2$

However, unlike undamped oscillator, the total energy of a damped oscillator is **not constant**, thus **it is not conserved**. The time rate of lost of total energy is given by $\frac{dH}{dt}$.

In[69]:= (* Lost rate of total energy *)
lostRateEnergy = D[H, t] // Simplify

Out[69]= $x'[t] \, (k \, x[t] + m \, x''[t])$

From the differential equation of damped harmonic oscillator above, we notice that the terms $k \, x[t] + m \, x''[t]$ equals to $-\gamma \, x'[t]$. So, by replacing it in the lost rate of total energy, we obtain

In[70]:= lostRateEnergy = lostRateEnergy /. (k x[t] + m x''[t]) → -γ x'[t]

Out[70]= $-\gamma \, x'[t]^2$

We notice here that the time rate of lost of total energy is due to the damping factor and is proportional to the square of velocity. With the presence of damping force, its value is always negative, thus the lost of total energy. If there is no damping force, γ is zero, then the total energy is constant just like the undamped oscillator.

Let us now compute the specific energies of the damped oscillator and plot the total energy and its time rate of lost energy versus time

Chapter 2. Linear Oscillatory Systems

In[71]:= `(* Kinetic energy of the underdamped oscillator *)`
`K1 = K /. solution /. ω0² → β² + ω1² // PowerExpand`

Out[71]= $\frac{1}{2} m \left(e^{t(-\beta - i\,\omega 1)} (-\beta - i\,\omega 1)\, C[1] + e^{t(-\beta + i\,\omega 1)} (-\beta + i\,\omega 1)\, C[2] \right)^2$

In[72]:= `(* Potential energy of the underdamped oscillator *)`
`V1 = V /. solution /. ω0² → β² + ω1² // PowerExpand`

Out[72]= $\frac{1}{2} k \left(e^{t(-\beta - i\,\omega 1)}\, C[1] + e^{t(-\beta + i\,\omega 1)}\, C[2] \right)^2$

In[73]:= `(* Total energy of the underdamped oscillator *)`
`H1 = K1 + V1 // Simplify`

Out[73]= $\frac{1}{2} \left(e^{-2t(\beta + i\,\omega 1)} k \left(C[1] + e^{2 i t \omega 1} C[2] \right)^2 + m \left(e^{-t(\beta + i\,\omega 1)} (\beta + i\,\omega 1) C[1] + e^{-t\beta + i t \omega 1} (\beta - i\,\omega 1) C[2] \right)^2 \right)$

In[74]:= `(* Plot the Total Energy and its Lost Rate vs Time *)`

$\text{Plot}\Big[\text{Evaluate}\Big[\{H1, D[H1, t]\} \Big] /. \Big\{ m \to 1, k \to 1, \omega 1 \to 1, \beta \to \frac{1}{7}, C[1] \to 1, C[2] \to 1 \Big\},$

$\{t, 0, 15\}, \text{AxesLabel} \to \Big\{ \text{"t"}, \text{"}H_1 \text{ (blue)}, \frac{dH_1}{dt} \text{ (green)"} \Big\},$

$\text{PlotStyle} \to \{\text{Blue, Green}\}, \text{PlotRange} \to \text{All} \Big]$

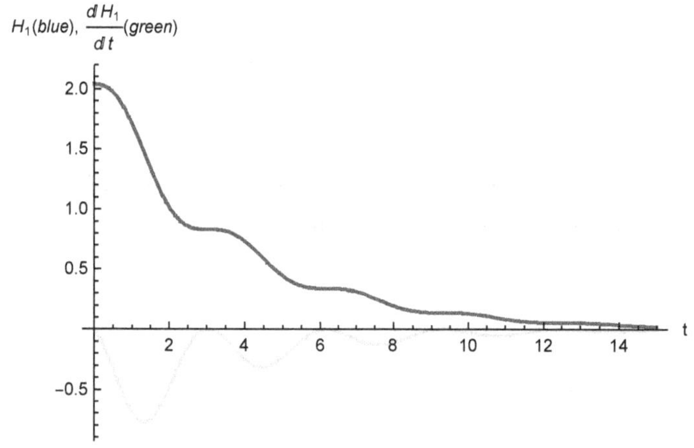

Figure 2.2.2 Total energy and its lost rate versus time in underdamped motion

As can be seen from **Figure 2.2.2**, both total energy and its rate of lost continue to gradually decrease until the oscillation dies out. The lost rate is maximum when the particle has maximum velocity, and it will instantaneously vanish when it has zero velocity, and continually the system is losing energy to the damping medium and is dissipated as heat.

2.2.1.2 Representation on the Phase Plane

Let us now investigate the phase path of the underdamped oscillator for a given initial conditions ($\omega_1 = 1$, $\beta = 1/7$, $C[1] = 1$, $C[2] = 1$) on the phase plane $(x, \dot{x})$.

In[75]:= `ParametricPlot[Evaluate[{underdampedSolution, D[underdampedSolution, t]} /.`
  `{ω1 → 1, β → `$\frac{1}{7}$`, C[1] → 1, C[2] → 1}], {t, 0, 25},`
  `AxesLabel → {"x", "`$\dot{x}$`"}, AspectRatio → Automatic, PlotRange → All]`

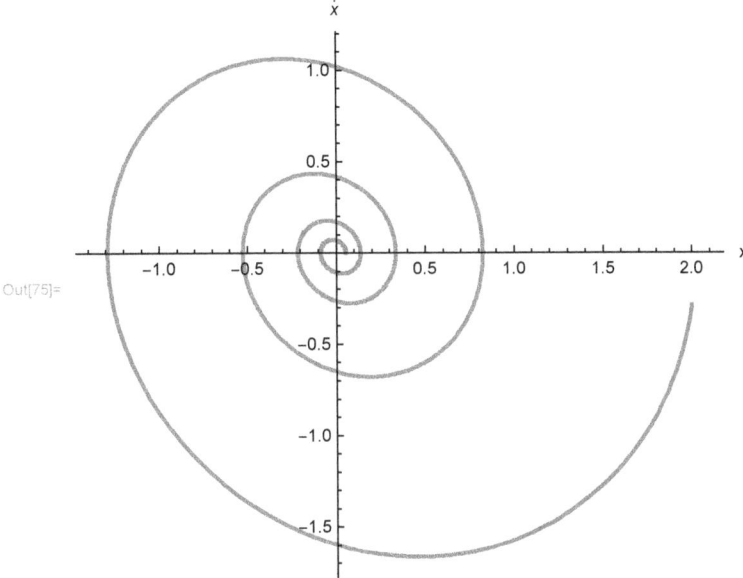

Figure 2.2.3 The phase path of the underdamped oscillator. The spiral path is the characteristic of the underdamped oscillator. ($\omega_1 = 1$, $\beta = 1/7$, $C[1] = 1$, $C[2] = 1$)

Figure 2.2.3 shows a spiral phase path of the underdamped oscillator. The continually decreasing magnitude of the radius vector for a representative point in the phase plane indicates the presence of damping force on the oscillator.

Here is another phase diagrams for various initial conditions

```
In[76]:= ParametricPlot[Evaluate[Table[Flatten[
    Refine[Re[ComplexExpand[{underdampedSolution, D[underdampedSolution, t]} /.
       {ω1 → 1, β → 1/7, C[1] → i, C[2] → 1}]], Element[t, Reals]]], {i, 1, 5}]],
    {t, 0, 25}, AxesLabel → {"x", "ẋ"}, AspectRatio → Automatic,
    PlotRange → All]
```

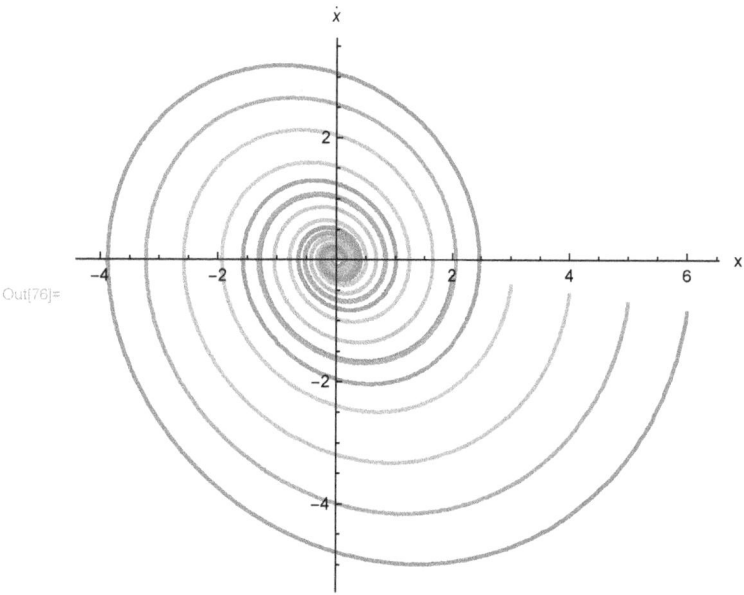

Figure 2.2.4 The Phase Diagram of the damped oscillator for various initial conditions ($\omega_1 = 1$, $\beta = 1/7$, $C[2] = 1$)

2.2.2 Case 2. Critically Damped Motion

If the damping force is large enough so that $\beta^2 > \omega_0^2$, the system is non-oscillatory. If the initial velocity is zero, the displacement decreases asymptotically from its initial value to the equilibrium position, $x = 0$.

The case of critical damping motion occurs when $\beta^2 = \omega_0^2$. The system is said to be in the boundary between overdamped and underdamped motion. Equation 2.2.2 gives the solution for the critically damped oscillator

```
In[77]:= criticallydampedSolution =
    x[t] /. Flatten[DSolve[{equation /. ω0^2 → β^2, x[0] == x0, x'[0] == v0}, x[t], t]]
Out[77]= e^(-t β) (t v0 + x0 + t x0 β)
```

where x_0 and v_0 are the initial values of the position and the velocity, respectively.

Let us plot the solution for a given initial conditions

```
In[78]:= criticallyDampedPlot =
    Plot[Evaluate[criticallydampedSolution /. {β → 1/5, x0 → 1, v0 → 0}],
     {t, 0, 25}, AxesLabel → {"t", "x"}]
```

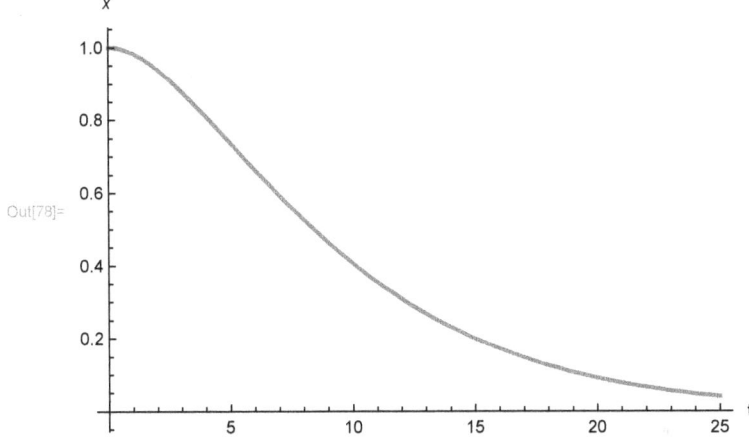

Out[78]=

Figure 2.2.5 The graph of solution for a critically damped oscillator $(\beta = 1/5,\ x(0) = 1,\ \dot{x}(0) = 0)$

2.2.2.1 Energy Considerations in Critically Damped Motion

The total energy of a critically damped motion

```
In[79]:= H1 = 1/2 k (criticallydampedSolution)^2 + 1/2 m (D[criticallydampedSolution, t])^2
```

Out[79]= $\frac{1}{2} e^{-2t\beta} k\,(t\,v0 + x0 + t\,x0\,\beta)^2 + \frac{1}{2} m\,\left(e^{-t\beta}\,(v0 + x0\,\beta) - e^{-t\beta}\,\beta\,(t\,v0 + x0 + t\,x0\,\beta)\right)^2$

```
(* Plot the Total Energy & Lost Rate vs Time *)
Plot[Evaluate[{H1, D[H1, t]} /. {m → 1, k → 1, β → 1/5, x0 → 1, v0 → 0}], {t, 0, 15},
 AxesLabel → {"t", "H₁ (blue), dH₁/dt (green)"}, PlotStyle → {Blue, Green}, PlotRange → All]
```

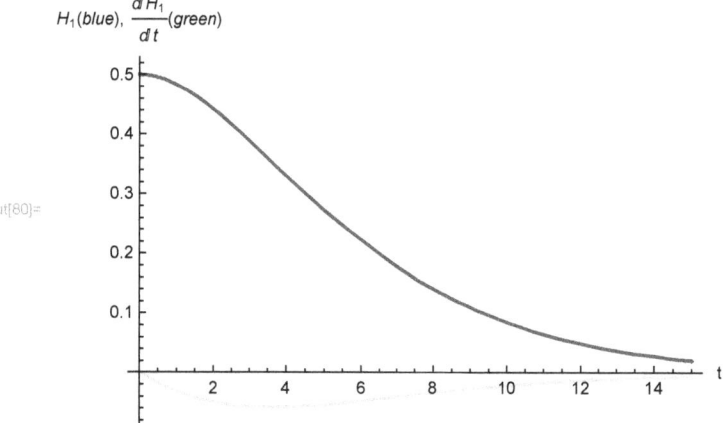

Figure 2.2.6 Total energy and its lost rate versus time in critically damped motion $(m = 1, k = 1, \beta = 1/5, x(0) = 1, \dot{x}(0) = 0)$

For a given set of initial conditions, the energy of a critically damped oscillator dies out more rapidly than that for an underdamped oscillator. This is an important requirement in designing certain oscillatory systems when the system must return to equilibrium as rapidly as possible.

2.2.2.2 Representation on the Phase Plane

The phase plane plot for motion starting off with the initial conditions $(x_0, v_0) = (1, 0)$ is

```
ParametricPlot[Evaluate[{criticallydampedSolution, D[criticallydampedSolution, t]} /.
  {β → 1/5, x0 → 1, v0 → 0}], {t, 0, 50},
 PlotRange → {{0, 1.1}, {-0.2, 0.2}}, AxesLabel → {"x", "ẋ"}]
```

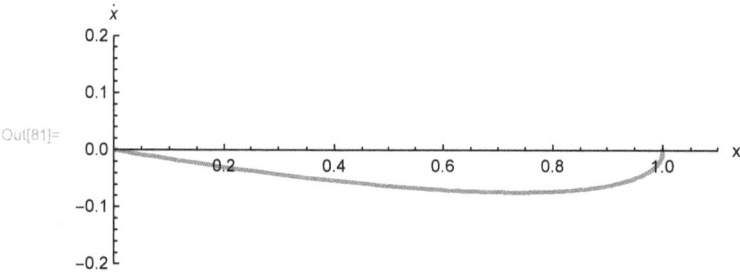

Figure 2.2.7 Phase-plane plot for a critically damped oscillator $(\beta = 1/5, x(0) = 1, \dot{x}(0) = 0)$

2.2.3 Case 3. Overdamped Motion

Overdamped motion occurs when β is larger than ω_0. Because $\beta^2 > \omega_0^2$, it is convenient to define $\omega_2^2 = \beta^2 - \omega_0^2$ where $\omega_2^2 > 0$. Thus, the exponents of the solution become real and the solution reduces to

In[82]:= `overdampedSolution = PowerExpand[x[t] /. solution /. ω0² → β² - ω2²]`

Out[82]= $e^{t(-\beta-\omega 2)} C[1] + e^{t(-\beta+\omega 2)} C[2]$

Note that
- ω_2 does not represent an angular frequency, because the motion is not periodic.
- the displacement approaches zero exponentially.

The two constants of integration C[1] and C[2] are related to initial values x_0 and v_0 as

In[83]:= `init1 = x0 == overdampedSolution /. t → 0`

Out[83]= $x0 == C[1] + C[2]$

In[84]:= `init2 = v0 == D[overdampedSolution, t] /. t → 0`

Out[84]= $v0 == (-\beta - \omega 2) C[1] + (-\beta + \omega 2) C[2]$

The initial values of C[1] and C[2] can be obtained by solving the initial equations involving the terms of x_0 and v_0. Thus,

$$x_0 = C[1] + C[2],$$
$$v_0 = (-\beta - \omega_2) C[1] + (-\beta + \omega_2) C[2].$$

In[85]:= `constants = Simplify[Solve[{init1, init2}, {C[1], C[2]}]]`

Out[85]= $\left\{\left\{C[1] \to -\frac{v0 + x0\,\beta - x0\,\omega 2}{2\,\omega 2},\ C[2] \to \frac{v0 + x0\,(\beta + \omega 2)}{2\,\omega 2}\right\}\right\}$

Substituting C[1] and C[2] back into the solution for the overdamped oscillator, it becomes

In[86]:= `overdampedSolution1 = overdampedSolution /. constants`

Out[86]= $\left\{-\frac{e^{t(-\beta-\omega 2)}(v0 + x0\,\beta - x0\,\omega 2)}{2\,\omega 2} + \frac{e^{t(-\beta+\omega 2)}(v0 + x0\,(\beta + \omega 2))}{2\,\omega 2}\right\}$

Let us plot the solution for a given set of conditions

In[87]:= `overDampedPlot = Plot[Evaluate[overdampedSolution1 /. {β → 1/5, ω2 → 1/10, x0 → 1, v0 → 0}], {t, 0, 25}, PlotStyle → Red, AxesLabel → {"t", "x"}, PlotRange → ( 0  25 / 0  1.1 )]`

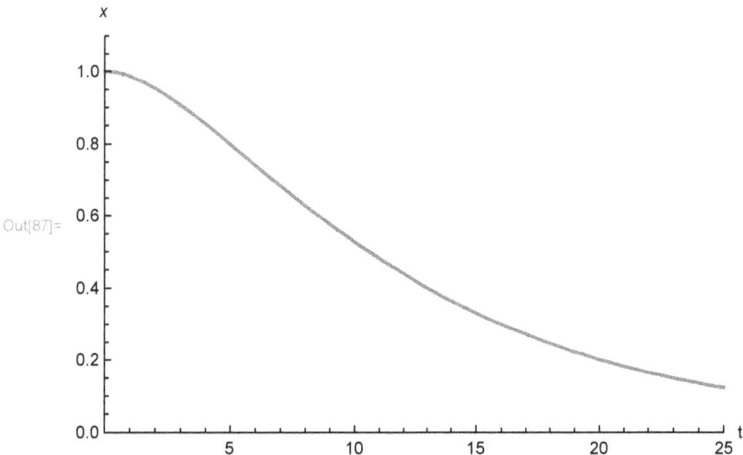

Figure 2.2.8 The graph of solution for the overdamped motion $\left(\beta = \frac{1}{5},\ \omega_2 = \frac{1}{10},\ x(0) = 1,\ \dot{x}(0) = 0\right)$

In[88]:= `(* Comparing the plot of the critically damped oscillator and the overdamped oscillator *)`
`Show[overDampedPlot, criticallyDampedPlot, Graphics[{Text[Overdamped, {10, 0.7}], Text[Critically damped, {5, 0.4}]}]]`

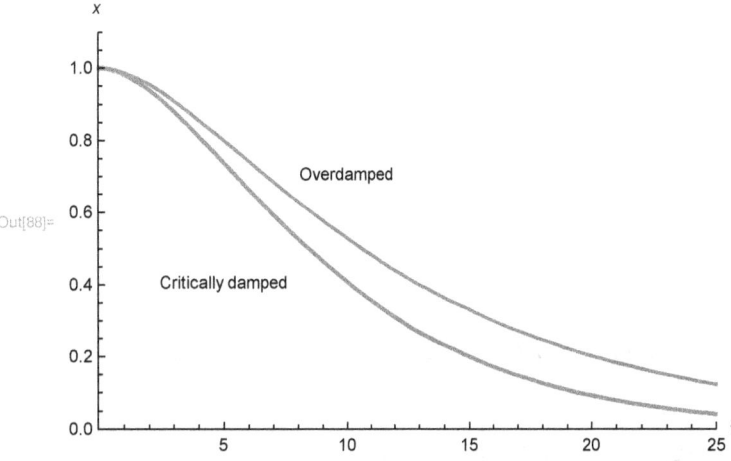

Figure 2.2.9 A graph of comparison between the solutions of a critically damped oscillator and an overdamped oscillator

For large time, the displacement approaches zero exponentially. However, the displacement in a decrease of amplitude may change sign before approaching zero. The following plots illustrate these possibilities. We shall limit our case where x(0) > 0 and specifically choose x(0) = 1 while changing v_0 accordingly

```
In[89]:= Plot[Evaluate[
    Table[overdampedSolution1 /. {β → 1/5, ω2 → 1/10, x0 → 1, v0 → v}, {v, -1, 1, 0.5}]],
  {t, 0, 25}, AxesLabel → {"t", "x"}, PlotRange → All]
```

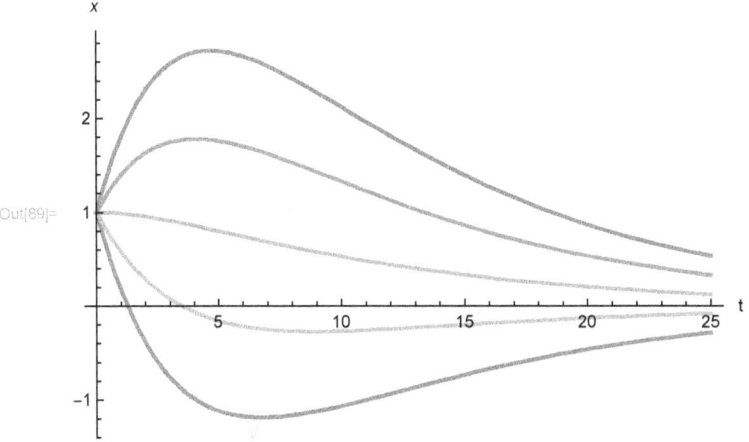

Figure 2.2.10 Displacement graphs for the solutions of overdamped motion with x(0) = 1 and with various initial velocities. Some curves change sign before approaching zero

Let us also plot the graph of various velocities vs time

```
In[90]:= Plot[Evaluate[
    Table[D[overdampedSolution1, t] /. {β → 1/5, ω2 → 1/10, x0 → 1, v0 → v}, {v, -1, 1, 0.5}]],
    {t, 0, 25}, AxesLabel → {"t", "ẋ"}, PlotRange → All]
```

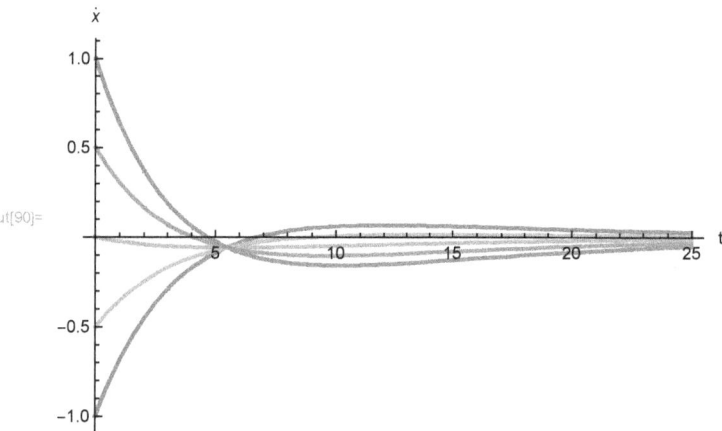

Out[90]=

Figure 2.2.11 The graph of velocity as a function of time for various initial velocities $\left(\beta = \frac{1}{5},\ \omega_2 = \frac{1}{10},\ x_0 = 1\right)$

Depending on the initial values of the position and the velocity, a change in sign of both x and $\dot{x}$ may occur as it approaches zero exponentially. We shall return to this again when we discuss the phase paths.

2.2.3.1 Energy Considerations in Overdamped Motion

Total energy of an overdamped motion

```
In[91]:= H1 = 1/2 k (overdampedSolution1)^2 + 1/2 m (D[overdampedSolution1, t])^2 // Simplify
```

$$\text{Out[91]}= \left\{ \frac{1}{8\,\omega 2^2} \left(k \left(e^{-t\,(\beta+\omega 2)} \,(v0 + x0\,(\beta - \omega 2)) - e^{t\,(-\beta+\omega 2)}\,(v0 + x0\,(\beta + \omega 2)) \right)^2 + m \left(e^{-t\,(\beta+\omega 2)}\,(v0 + x0\,(\beta - \omega 2))\,(\beta + \omega 2) - e^{t\,(-\beta+\omega 2)}\,(\beta - \omega 2)\,(v0 + x0\,(\beta + \omega 2)) \right)^2 \right) \right\}$$

In[92]:= (* Plot the total energy and its lost rate vs time *)
Plot[Evaluate[{H1, D[H1, t]} /. {m → 1, k → 1, ω2 → $\frac{1}{10}$, β → $\frac{1}{5}$, x0 → 1, v0 → 0}],
{t, 0, 15}, AxesLabel → {"t", "H₁(blue), $\frac{dH_1}{dt}$ (green)"},
PlotStyle → {Blue, Green}, PlotRange → All]

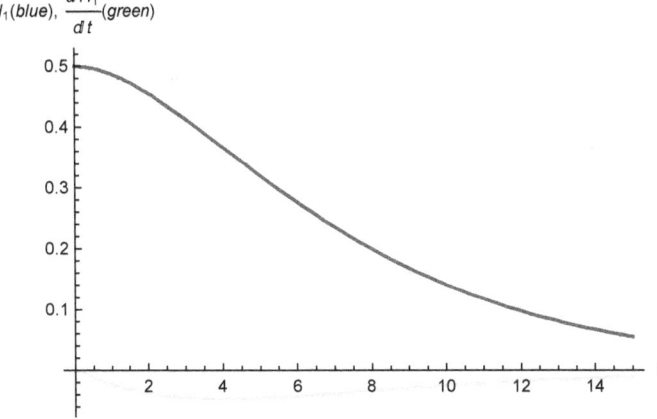

Figure 2.2.12 Total energy and its lost rate versus time in overdamped motion

Similar to the case in critically damped motion, the energy dies out exponentially. However, the energy of a critically damped oscillator dies out more rapidly than that of an overdamped oscillator.

2.2.3.2 Representation on the Phase Plane

The phase diagram is constructed by plotting x versus $\dot{x}$.

In[93]:= (* Phase diagram for initial conditions: x0=1 and v0=0
*)
ParametricPlot[Evaluate[Flatten[{overdampedSolution1, D[overdampedSolution1, t]}]] /.
{β → $\frac{1}{5}$, ω2 → $\frac{1}{10}$, x0 → 1, v0 → 0}], {t, 0, 25},
PlotRange → {{0, 1.1}, {-0.2, 0.2}}, AxesLabel → {"x", "$\dot{x}$"}]

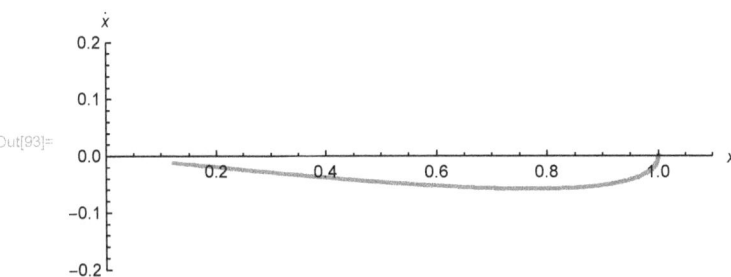

Figure 2.2.13 The phase diagram of an overdamped motion $\left(\beta = \frac{1}{5},\ \omega_2 = \frac{1}{10},\ x_0 = 1,\ \dot{x}_0 = 0\right)$

Overdamping results may exhibit some strange behavior as can be seen in the phase space diagram if we plot it for various values of the initial velocity

In[94]:= `overdampedPlots = ParametricPlot[`
   `Evaluate[Table[Flatten[{overdampedSolution1, D[overdampedSolution1, t]}] /.`
   $\left\{\beta \to \frac{1}{5},\ \omega 2 \to \frac{1}{10},\ x0 \to 1,\ v0 \to v\right\},\ \{v, -1, 1, 0.25\}]\],$
   `{t, 0, 25}, AxesLabel → {"x", "`$\dot{x}$`"}, PlotRange → All]`

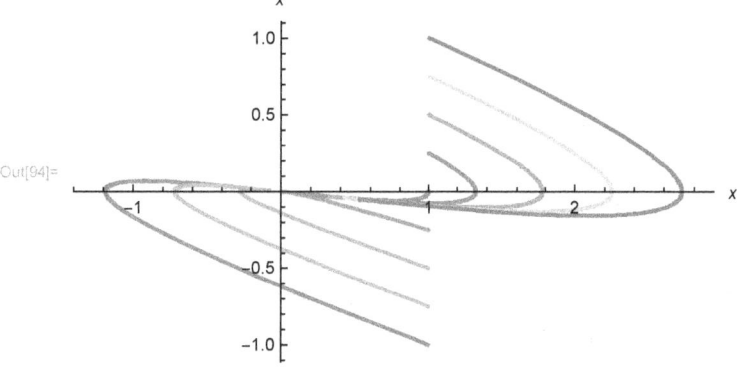

Figure 2.2.14. The phase paths for overdamped motion are shown for initial displacement $x_0 = 1$ with various initial values $\dot{x}_0 = v_0$. $\left(\beta = \frac{1}{5},\ \omega_2 = \frac{1}{10}\right)$

Note that at longer time for all the phase paths of the various initial velocities shown, there corresponds an asymptotic line $\dot{x} = -(\beta - \omega 2)\,x$ (see S. T. Thornton, J. B. Marion, Classical Dynamics of Particles and Systems 5th edition)

```
In[95]:= overdampedAsymptote = Plot[Evaluate[{-(β - ω2) x, -(β + ω2) x} /. {β → 1/5, ω2 → 1/10}],
    {x, -1.5, 3}, PlotStyle → {{Dashed, Red}, {Dashed, Blue}}]
```

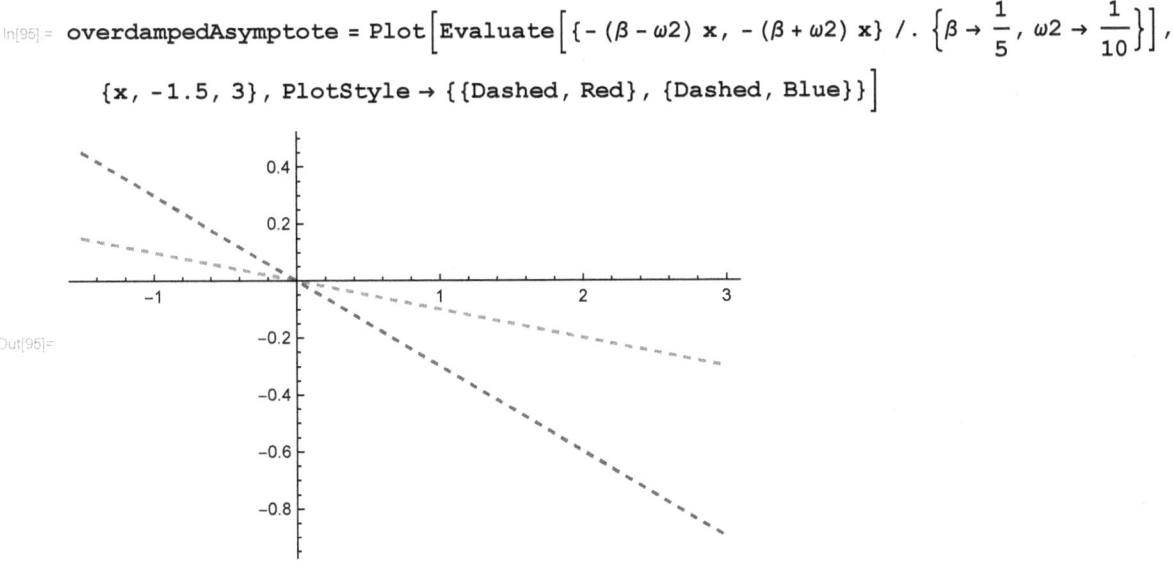

Figure 2.2.15 Asymptotic lines corresponding to the phase paths of overdamped motion

```
In[96]:= (* Asymptotic line: x'=-(β-ω₂)x and x'=-(β+ω₂)x *)
    Show[overdampedPlots, overdampedAsymptote,
     Graphics[{Text["ẋ=-(β-ω₂)x", {2.5, -0.4}],
       Text["ẋ=-(β+ω₂)x", {2.2, -0.9}], Text["I", {0.8, 0.5}],
       Text["II", {0.7, 0.07}], Text["III", {1.2, -0.7}]}], AspectRatio → 1]
```

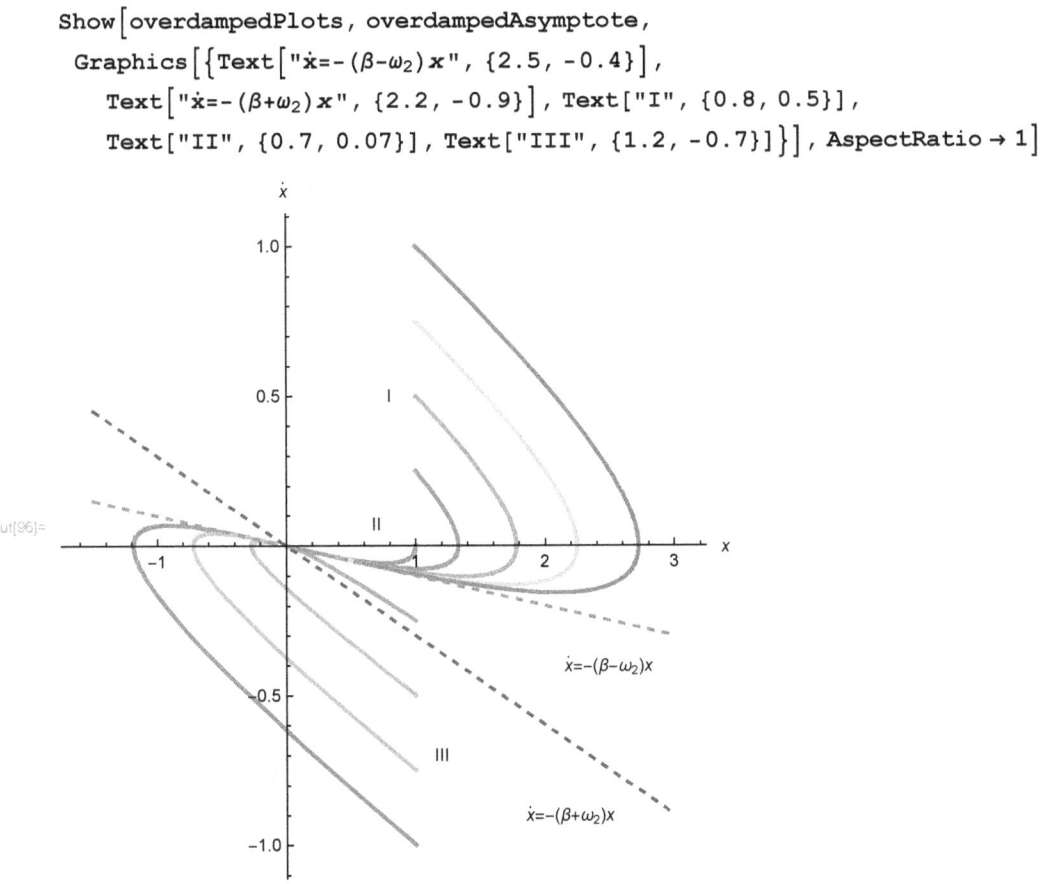

Figure 2.2.16. The graph displays x and $\dot{x}$ as a function of time for three regions of phase paths labeled I, II, and III, superimposed with asymptotic lines $\dot{x} = -(β - ω_2) x$ and $\dot{x} = -(β + ω_2) x$

Depending on the initial values of the position and the velocity, change in signs of both x and $\dot{x}$ may occur; for example, with a given positive initial displacement $x(0) = x_0 = 1$, there are three cases of interest for various initial velocities $\dot{x}(0) = v_0$.

Region I. $v_0 > 0$

$x(t)$ reaches maximum at some $t > 0$; decreases, becomes negative and then approaches zero.

Region II. $v_0 \leq 0$

$x(t)$ and $\dot{x}(t)$ decrease monotonically while approaching zero.

Note that there is a phase path whose initial point lying between the two dashed curves in **Figure 2.2.16** which is decreasing monotonically to zero.

Region III. $v_0 < 0$

v_0 is below the curve $x' = -(\beta + \omega_2) x$; $x(t)$ is positive but goes negative before reaches a minimum value, and then approaches zero. $\dot{x}(t)$ is negative but goes positive before approaching zero. The motion in this case could be considered oscillatory.

2.3 Sinusoidal Driven Oscillations

In this section, we study the damped harmonic oscillations in which an external sinusoidal driving force is applied to the oscillator.

Suppose a force of the form $F_0 \cos(\omega t)$ is exerted upon such an oscillator. The total force on the particle is then

(2.3.1) $$F = -kx - \gamma \dot{x} + F_0 \cos(\omega t),$$

where $(-kx)$ is a linear restoring force and $(-\gamma \dot{x})$, a viscous damping force.

The equation of motion becomes

(2.3.2) $$m \frac{d^2 x}{dt^2} = -kx - \gamma \dot{x} + F_0 \cos(\omega t)$$

```
In[97]:= Clear["Global`*"]

In[98]:= (* Equation 2.3.2 *)
        equationDriven = m D[x[t], {t, 2}] == -k x[t] - γ D[x[t], t] + F0 Cos[ω t]

Out[98]= m x''[t] == F0 Cos[t ω] - k x[t] - γ x'[t]
```

Using our previous definitions, $\beta = \dfrac{\gamma}{2m}$ (damping factor) and $\omega_0 = \sqrt{\dfrac{k}{m}}$ (the natural frequency in the absence of damping), the equation of motion can be re-written as

$$\frac{d^2x}{dt^2} + \omega_0^2\, x + 2\beta\, \dot{x} = F_0 \cos(\omega t),$$

In[99]:= `equationDriven = D[x[t], {t, 2}] + w0^2 x[t] + 2 β D[x[t], t] == A Cos[ω t]`

Out[99]= $\omega 0^2\, x[t] + 2\beta\, x'[t] + x''[t] == A\, \text{Cos}[t\,\omega]$

where $A \equiv F_0/m$ is the **reduced amplitude** and ω is the frequency of the driving force respectively. The solutions of this equation is

In[100]:= `solutionDriven = DSolve[equationDriven, x[t], t] // Simplify`

Out[100]= $\left\{\left\{x[t] \to e^{-t\left(\beta+\sqrt{\beta^2-\omega 0^2}\right)} C[1] + e^{t\left(-\beta+\sqrt{\beta^2-\omega 0^2}\right)} C[2] + \dfrac{A\,((-\omega^2+\omega 0^2)\,\text{Cos}[t\,\omega] + 2\beta\,\omega\,\text{Sin}[t\,\omega])}{4\beta^2\,\omega^2 + (\omega^2-\omega 0^2)^2}\right\}\right\}$

We observe that the solution consists of two parts:

1) The first part represents the **complementary solution** (x_c) containing initial conditions denoted by the constants of integration C[1] and C[2]. It is independent of driving force.

$$e^{-t\left(\beta+\sqrt{\beta^2-\omega 0^2}\right)} C[1] + e^{t\left(-\beta+\sqrt{\beta^2-\omega 0^2}\right)} C[2]$$

This is the solution of equation 2.3.2 with F_0 set equal to zero. So, it is a solution for damped oscillator as in Section 2.2.

2) The second part is the **particular solution** (x_p) which is free of any constant of integration. This part is present in any case independent of the initial conditions.

$$\left(A\,((-\omega^2+\omega 0^2)\,\text{Cos}[t\,\omega] + 2\beta\,\omega\,\text{Sin}[t\,\omega])\right) \Big/ \left(4\beta^2\,\omega^2 + (\omega^2-\omega 0^2)^2\right)$$

Thus, the general solution can be written as

$$x(t) = x_c(t) + x_p(t).$$

In[101]:= `particularSolution = x[t] /. solutionDriven /. {C[1] → 0, C[2] → 0}`

Out[101]= $\left\{\dfrac{A\,((-\omega^2+\omega 0^2)\,\text{Cos}[t\,\omega] + 2\beta\,\omega\,\text{Sin}[t\,\omega])}{4\beta^2\,\omega^2 + (\omega^2-\omega 0^2)^2}\right\}$

In[102]:= `complementarySolution = (x[t] /. solutionDriven) - particularSolution`

Out[102]= $\left\{ e^{-t\left(\beta+\sqrt{\beta^2-\omega 0^2}\right)} C[1] + e^{t\left(-\beta+\sqrt{\beta^2-\omega 0^2}\right)} C[2] \right\}$

Let us plot the solutions for a set of given conditions

In[103]:= `Plot[complementarySolution /. {C[1] → 1, C[2] → 1, ω0 → 1, β → 0.2},`
`{t, 0, 10 Pi}, PlotRange → All]`

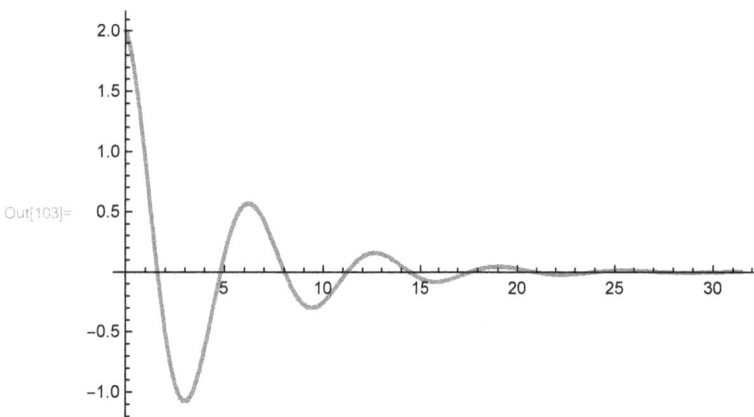

Figure 2.3.1. Graphic plot of a complementary solution of the driven, damped oscillator ($C[1] = 1$, $C[2] = 1$, $\omega_0 = 1$, $\beta = 0.2$)

The complementary solution here represents **transient effects** (that is, effects that die out), and the curve damp out with time because of the factor $e^{-\beta t}$.

In[104]:= `Plot[particularSolution /. {A → 1, ω0 → 1, β → .2, ω → 2}, {t, 0, 10 Pi}]`

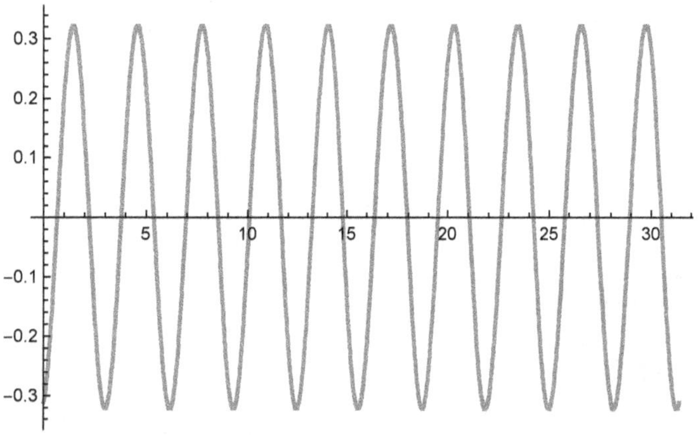

Figure 2.3.2. Graphic plot of a particular solution of the driven, damped oscillator ($A \equiv F_0/m = 1$, $\omega_0 = 1$, $\beta = 0.2$, $\omega = 2$)

The particular solution represents the **steady-state effects**. The steady-state solution is important in many applications and problems.

Let us now plot the whole solution

```
In[105]:= Plot[x[t] /. solutionDriven /.
          {C[1] → 1, C[2] → 1, A → 1, ω0 → 1, β → .2, ω → 2}, {t, 0, 10 π}]
```

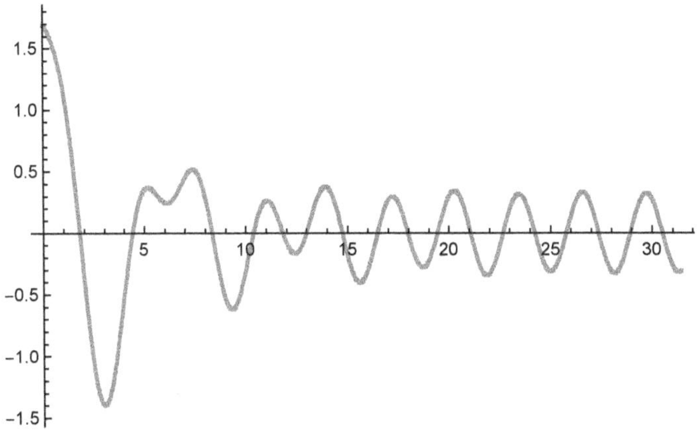

Figure 2.3.3. Graphic plot of the driven, damped oscillator. The driving force causes the oscillator to vibrate at the driving frequency ω at large t ($C[1] = 1$, $C[2] = 1$, $A = 1$, $\omega_0 = 1$, $\beta = 0.2$, $\omega = 2$)

As t large compared with $\frac{1}{\beta}$, the steady-state solution takes charge, $x\left(t \geq \frac{1}{\beta}\right) = x_p$, where x_p is the steady-state solution.

The details of the motion while it is in the transient period (that is., $t \leq \frac{1}{\beta}$) is dependent on the oscillator's conditions at the time that the driving force is first applied and also on the driving frequency, ω. Let us show this in the following graph

Chapter 2. Linear Oscillatory Systems

```
In[106]:= Show[Plot[x[t] /. solutionDriven /. {C[1] → 1, C[2] → 1, A → 1, ω0 → 1, β → .15, ω → 1/7},
   {t, 0, 30 π}, Epilog → {Text[ω₀ > ω, {8, 1.65}], Text[ω₀ < ω, {8, -1}]},
   PlotStyle → {Red, Thickness[0.005]}, PlotRange → All, AxesLabel → {"t", "ẋ"}],
 Plot[x[t] /. solutionDriven /. {C[1] → 1, C[2] → 1, A → 1, ω0 → 1, β → .3, ω → 5},
   {t, 0, 30 π}, PlotRange → All]]
```

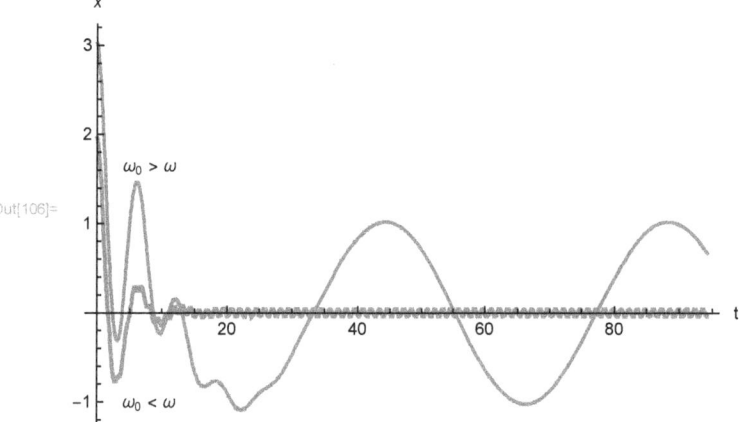

Figure 2.3.4. Comparisons of two sinusoidal driven oscillatory motion with damping, for $\omega_0 > \omega$ and $\omega_0 < \omega$

Note that for both motions, the initial transient periods are different. Eventually, however, both systems will be governed by the steady-state effects at larger time.

2.3.1 Resonance Phenomena

The particular solution x_p represents the **steady state effects** and contains all of the information about the system for large t. For this reason we shall first examine the particular solution. The important part of the particular solution x_p, is its amplitude.

At $t = 0$, the initial amplitude is

```
In[107]:= amplitudeInitialP = particularSolution /. t → 0
```

$$\text{Out[107]}= \left\{ \frac{A\,(-\omega^2 + \omega 0^2)}{4\,\beta^2\,\omega^2 + (\omega^2 - \omega 0^2)^2} \right\}$$

The following graph shows the relation between initial amplitude with the frequency of the driving force ω for various values of the damping factor β.

In[108]:= `Plot[Evaluate[Table[amplitudeInitialP /. {A → 1, ω0 → 1, β → i}, {i, 0.1, 1.2, 0.2}]],`
`{ω, 0, 3}, AxesLabel → {"ω", "Amplitude of x_p"}]`

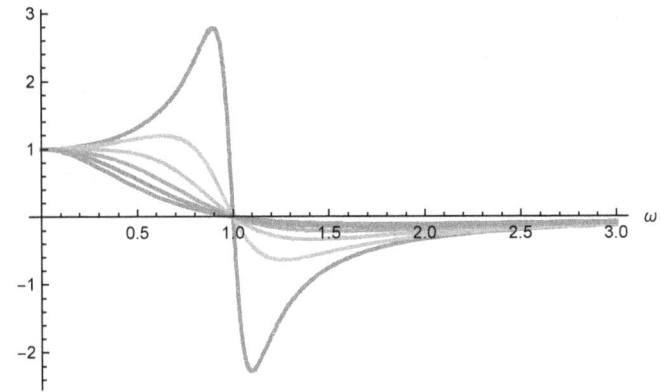

Out[108]=

Figure 2.3.5. Graph showing relation of initial amplitude with the frequency of the driving force ω for various values of the damping factor β ($A = 1$, $\omega_0 = 1$).

The initial amplitude become large, especially if the damping factor β is small as ω approaches ω_0. In present case $\omega_0 = 1$.

To find the amplitude of our particular solution, we notice that it can be re-written as $x_p = d \cos(\omega t - \alpha)$, where d is the amplitude and α is the phase shift factor. The phase shift occurs because the driving frequency ω is different from the natural frequency of the non-driven oscillator ω_0.

By using the trigonometric addition formulas, we identify the following equations

(2.3.3) $$d \cos \alpha = \frac{A(\omega_0^2 - \omega^2)}{4\beta^2 \omega^2 + (\omega^2 - \omega_0^2)^2}$$

(2.3.4) $$d \sin \alpha = \frac{2A\beta\omega}{4\beta^2 \omega^2 + (\omega^2 - \omega_0^2)^2}$$

By dividing equation 2.3.4 over equation 2.3.3 we obtain the phase shift as

In[109]:= $\alpha = \text{ArcTan}\left[\dfrac{2\beta\omega}{(\omega0^2 - \omega^2)}\right];$

By squaring the above equations (2.3.3 and 2.3.4) and using relations among trigonometric functions, we obtain the amplitude d of our steady-state solution as

In[110]:= $\text{amplitudeP} = \dfrac{A}{\sqrt{4\beta^2 \omega^2 + (\omega^2 - \omega0^2)^2}}$

Out[110]= $\dfrac{A}{\sqrt{4\beta^2 \omega^2 + (\omega^2 - \omega0^2)^2}}$

The result shows that both the amplitude and phase shift of the particular solution depends on the driving frequency ω, the damping factor β, the natural frequency of the oscillator ω_0, and for the amplitude, on the reduced amplitude of the applied driving force A.

There is an interesting situation, when the driving frequency ω matches the natural frequency ω_0. In such case, which is referred to as a **resonance**, the amplitude can become large, especially if the damping factor β is small.

To find the resonance frequency ω_R at which the amplitude is maximum, we set the derivative of the amplitude with respect to ω equal zero, solve the resulting equation for ω, and use **resonanceAmplitude** to denote the solution

In[111]:= `resonanceAmplitude = Solve[D[amplitudeP, ω] == 0, ω]`

Out[111]= $\left\{\{\omega \to 0\}, \left\{\omega \to -\sqrt{-2\beta^2 + \omega 0^2}\right\}, \left\{\omega \to \sqrt{-2\beta^2 + \omega 0^2}\right\}\right\}$

Since the frequency has to be positive and nontrivial, we choose the resonance frequency ω_R to be

In[112]:= `ωR = ω /. resonanceAmplitude[[3]]`

Out[112]= $\sqrt{-2\beta^2 + \omega 0^2}$

Note that:

1. The resonance frequency ω_R becomes smaller as the damping coefficient β is increased.
2. ω_R is imaginary if $\omega_0^2 - 2\beta^2 < 0$.

Thus, at strong damping case, no resonance occurs if $\beta > \dfrac{\omega_0}{\sqrt{2}}$, for then the amplitude d decreases monotonically with increasing ω. To see this, consider the limiting case $\beta = \dfrac{\omega_0}{\sqrt{2}}$,

In[113]:= `amplitudeP /. β → ω0/√2 // Simplify`

Out[113]= $\dfrac{A}{\sqrt{\omega^4 + \omega 0^4}}$

which clearly decreases with increasing values of ω, starting with $\omega = 0$.

We can now plot the amplitude for various values of damping factor β.

In[114]:= `(* Amplitude for various values of β *)`
`Plot[Evaluate[Table[amplitudeP /. {A → 1, ω0 → 1, β → i}, {i, 0.1, 1.2, 0.2}]],`
`  {ω, 0, 2}, AxesLabel → {"ω", "Amplitude"}, PlotRange → {{0, 2}, {0, 6}}]`

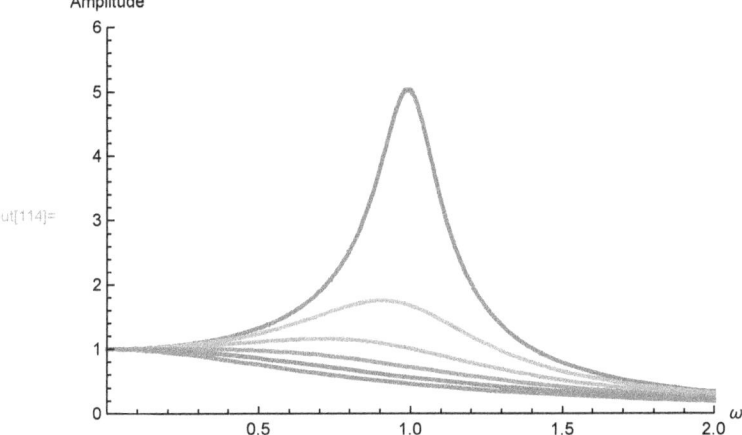

Figure 2.3.6. The amplitude of the steady-state solution is displayed as a function of the driving frequency ω for various values of β.

We observe that if the driving frequency is approximately equal to the natural frequency of the system ($\omega_0 = 1$ in present case), the system will oscillate with a very large amplitude. This phenomenon is called **amplitude resonance**.

The amplitude at the resonance peak of our steady-state oscillation can be computed as follows

In[115]:= `amplitudePMax = amplitudeP /. ω → `$\sqrt{\omega 0^2 - 2 \beta^2}$` // Factor`

Out[115]= $\dfrac{A}{2\sqrt{-\beta^2(\beta^2 - \omega 0^2)}}$

Notice from the above, the smaller the value of the damping factor β, the bigger is the amplitude at the resonance peak.

The graphical representation of the phase shift for various values of the damping constant is given below

```
(** Phase shift for various values of β **)
Plot[Evaluate[Table[α /. {ω0 → 1, β → i}, {i, 0.1, 1.2, 0.2}]],
 {ω, 0, 4}, AxesLabel → {"ω", "α"}, PlotRange → {{0, 4}, {-2, 2}},
 Exclusions → ω == 1, ExclusionsStyle → Directive[Red, Dashed]]
```

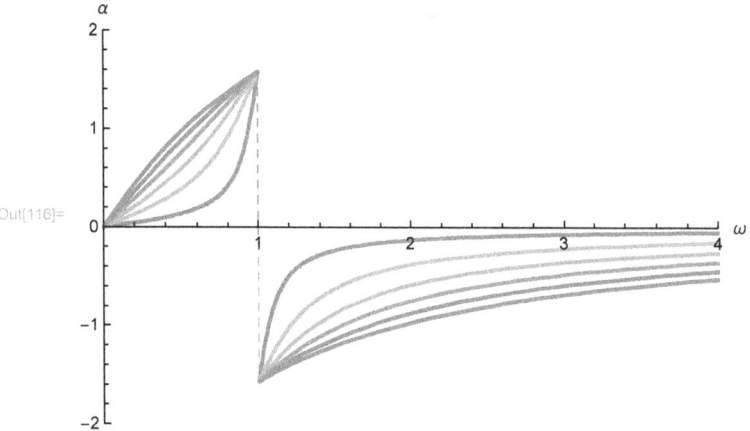

Figure 2.3.7. Phase shift for different values of damping factor, β. The phase shift changes discontinuously at $\omega = \omega_0$ where $\alpha = \frac{\pi}{2}$.

The phase shift α gives the difference in phase α between the applied driving force and the steady-state response. There is a real delay between the action of the driving force and the response of the system.

Let us now summarize the oscillation frequencies for all the different cases we have considered so far:

1. Free oscillations, no damping (Section 2.1)

$$\omega_0 = \sqrt{\frac{k}{m}}$$

2. Free oscillations with damping (Section 2.2)

    a. Underdamped motion
$$\omega_1^2 = \omega_0^2 - \beta^2, \text{ where } \omega_1^2 > 0.$$

    b. Critically damped motion
$$\omega_0^2 = \beta^2$$

    c. Overdamped motion
$$\omega_2^2 = \beta^2 - \omega_0^2 \text{ where } \omega_2^2 > 0.$$

3. Driven oscillations with damping

$$\omega_R^2 = \omega_0^2 - 2\beta^2$$

Note that $\omega_0 > \omega_1 > \omega_R$.

2.3.2 Energy Considerations in Resonance

Similar to amplitude resonance, energy resonance occurs when the kinetic energy becomes a maximum. Since the transient state dies out for large t, the kinetic energy is governed by the steady-state solution. In such condition, the kinetic energy becomes

In[117]:= `K = 1/2 m (D[particularSolution, t])^2 // Simplify`

Out[117]= $\left\{ \dfrac{A^2\, m\, \omega^2\, (2\,\beta\,\omega\, \cos[t\,\omega] + (\omega^2 - \omega0^2)\, \sin[t\,\omega])^2}{2\, (4\,\beta^2\,\omega^2 + (\omega^2 - \omega0^2)^2)^2} \right\}$

To obtain a value of kinetic energy independent of time, we compute the average of K over one complete period of oscillation defined as

$$\langle K \rangle \equiv \frac{1}{T} \int_0^T K\, dt,$$

where $T = \dfrac{2\pi}{\omega}$.

In[118]:= `⟨K⟩ = ω/(2π) ∫_0^(2π/ω) K dt`

Out[118]= $\left\{ \dfrac{A^2\, m\, \omega^2}{4\, (4\,\beta^2\,\omega^2 + (\omega^2 - \omega0^2)^2)} \right\}$

Similar to our previous procedure in finding ω_R, we find the resonance frequency ω_E at which the kinetic energy is maximum, by setting the derivative of the kinetic energy with respect to ω equal zero, and solve the resulting equation for ω

In[119]:= `resonanceKinetic = Solve[D[⟨K⟩, ω] == 0, ω]`

Out[119]= $\{\{\omega \to 0\}, \{\omega \to -\omega0\}, \{\omega \to -i\,\omega0\}, \{\omega \to i\,\omega0\}, \{\omega \to \omega0\}\}$

Since trivial, negative, and complex solutions have less physical importance, we choose

In[120]:= `ωE = ω /. resonanceKinetic[[5]]`

Out[120]= $\omega0$

Thus, the kinetic energy resonance occurs at the natural frequency of the system, ω_0.

Let us now find the potential energy resonance of the oscillator. Similar to our argument for the kinetic energy, the potential energy is governed by the steady-state solution as

In[121]:= `V = 1/2 m (particularSolution)^2 // Simplify`

Out[121]= $\left\{ \dfrac{A^2 \, m \, ((-\omega^2 + \omega 0^2) \, \text{Cos}[t\,\omega] + 2\,\beta\,\omega\,\text{Sin}[t\,\omega])^2}{2 \left(4\,\beta^2\,\omega^2 + (\omega^2 - \omega 0^2)^2\right)^2} \right\}$

The average of the potential energy over one complete period of oscillation is defined as

$$\langle V \rangle \equiv \frac{1}{T} \int_0^T V \, dt,$$

where $T = \dfrac{2\pi}{\omega}$.

In[122]:= $\langle V \rangle$ = `ω/(2 π) Integrate[V, {t, 0, 2π/ω}]`

Out[122]= $\left\{ \dfrac{A^2 \, m}{4 \left(4\,\beta^2\,\omega^2 + (\omega^2 - \omega 0^2)^2\right)} \right\}$

In[123]:= `resonancePotential = Solve[D[⟨V⟩, ω] == 0, ω]`

Out[123]= $\left\{ \{\omega \to 0\},\ \left\{\omega \to -\sqrt{-2\,\beta^2 + \omega 0^2}\right\},\ \left\{\omega \to \sqrt{-2\,\beta^2 + \omega 0^2}\right\} \right\}$

Choosing the nontrivial and positive solution, the potential energy resonance also occurs at $\sqrt{\omega_0^2 - 2\beta^2}$, just as the amplitude. It does not come as a surprise though, since the potential energy is proportional to the square of the amplitude.

We see therefore that the kinetic and potential energies resonate at different frequencies. This is a result of the fact that the damped and driven oscillator is not a conservative system. Energy is continually exchanged with the driving mechanism, and being transferred to the damping medium.

2.3.3 Representation on the Phase Plane

Let us plot the phase path of the steady-state solution of our driven, damped oscillator for various initial conditions on the phase plane $(x, \dot{x})$.

```
In[124]:= ParametricPlot[
    Evaluate[Table[Flatten[{particularSolution, D[particularSolution, t]}] /.
        {A → i, ω0 → 1, β → .2, ω → 0.5}, {i, 0, 2, 0.5}]],
    {t, 0, 50}, AxesLabel → {"x", "ẋ"}, PlotRange → All]
```

Out[124]=

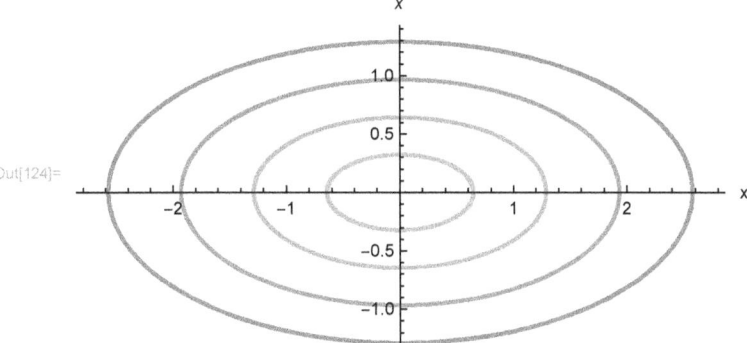

Figure 2.3.8. A complete phase diagram of a driven, damped oscillator for various initial conditions ($\omega_0 = 1$, $\beta = 0.2$, $\omega = 0.5$)

We expect the phase diagram to be similar with the one in simple harmonic oscillator, since the steady-state solution at large time is governed by the frequency of the driving force, ω.

2.3.4 Impulsive Forcing Functions

Suppose now that the driving force is discontinuous such that it is represented by a step force function or an impulse force function as shown in Figure 2.3.9 below

The step force function H is given by

$$H(t_0) = \begin{cases} 0, & t < t_0 \\ a, & t > t_0, \end{cases}$$

where a is constant and the argument t_0 indicates that the time of application of the force is at $t = t_0$.

Such function is represented by **HeavisideTheta function** in *Mathematica*,

```
In[125]:= heaviside[t_, t0_] := a HeavisideTheta[t - t0]
```

Let us plot the graph of the step function

In[126]:= `Plot[heaviside[t, 1] /. a → 1, {t, -2, 2}]`

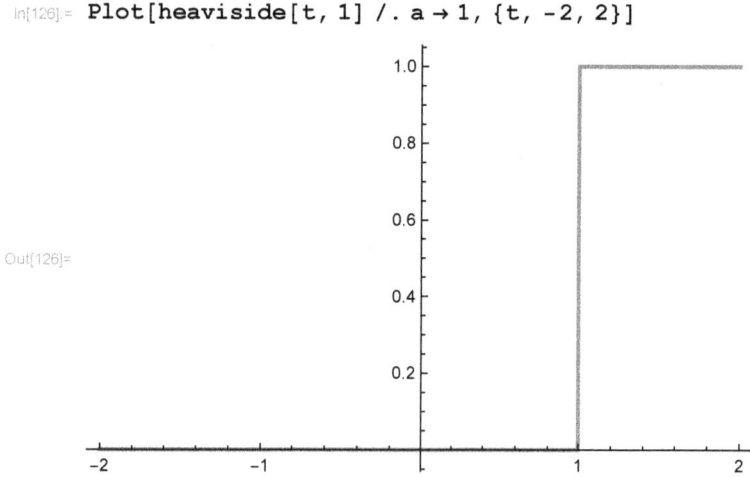

Figure 2.3.9 The graph of a step force function

The impulse force function $\mathcal{I}$ is a positive step function $H(t_0)$ applied at $t = t_0$, followed by a negative step function $-H(t_1)$ applied at $t = t_1$,

thus,

$$\mathcal{I}(t_0, t_1) = H(t_0) - H(t_1)$$

$$\mathcal{I}(t_0, t_1) = \begin{cases} 0, & t < t_0 \\ a, & t_0 < t < t_1 \\ 0, & t > t_1. \end{cases}$$

In[127]:= `impulse[t_, t0_, t1_] := a (HeavisideTheta[t - t0] - HeavisideTheta[t - t1])`

In[128]:= `Plot[impulse[t, 1, 3] /. a → 1, {t, -2, 4}]`

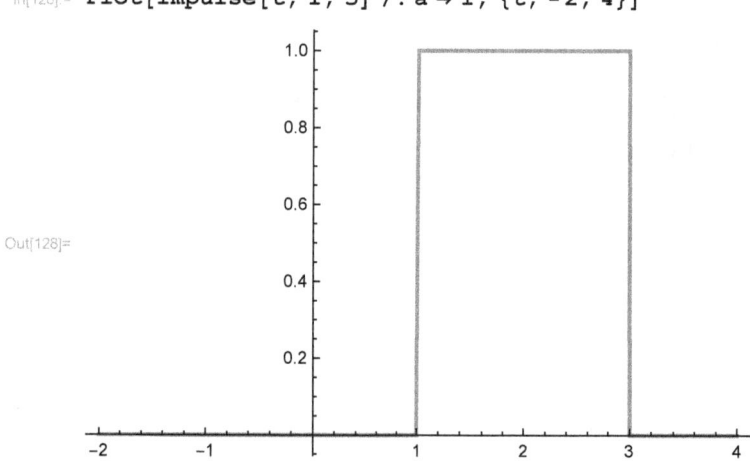

Figure 2.3.10 The graph of an impulse force function

2.3.4.1 Response to a Force Step Function

For step force functions, the equation of motion becomes

$$\frac{d^2x}{dt^2} + \omega_0^2 x + 2\beta \dot{x} = a, \qquad \text{for } t > t_0.$$

In[129]:= `equationStepDriven = D[x[t], {t, 2}] + w0^2 x[t] + 2 β D[x[t], t] == a`

Out[129]= $\omega 0^2\, x[t] + 2\,\beta\, x'[t] + x''[t] == a$

We consider $t_0 = 0$, so the solution of this equation for $t > 0$ is

In[130]:= `solutionStepDriven = DSolve[equationStepDriven, x[t], t] // Simplify`

Out[130]= $\left\{\left\{x[t] \to \dfrac{a + e^{-t\left(\beta+\sqrt{\beta^2-\omega 0^2}\right)}\,\omega 0^2\left(C[1] + e^{2t\sqrt{\beta^2-\omega 0^2}}\,C[2]\right)}{\omega 0^2}\right\}\right\}$

In[131]:= `particularStepSolution = x[t] /. solutionStepDriven /. {C[1] → 0, C[2] → 0}`

Out[131]= $\left\{\dfrac{a}{\omega 0^2}\right\}$

The particular solution $x_p(t)$ is just a constant: $\dfrac{a}{\omega 0^2}$

In[132]:= `complementaryStepSolution =`
`    (x[t] /. solutionStepDriven) - particularStepSolution // Simplify`

Out[132]= $\left\{e^{-t\left(\beta+\sqrt{\beta^2-\omega 0^2}\right)}\left(C[1] + e^{2t\sqrt{\beta^2-\omega 0^2}}\,C[2]\right)\right\}$

The complementary solution $x_c(t)$ is the same with the one in sinusoidal driving force.

Thus, the general solution can be written as

$$x(t) = x_c(t) + x_p(t).$$

Let us consider the initial conditions to be $x(t_0 = 0) = 0$, and $\dot{x}(t_0 = 0) = 0$, and find the constants of integration C[1] and C[2]

```
In[133]:= constants =
    Solve[{(x[t] /. solutionStepDriven) == 0, (D[x[t] /. solutionStepDriven, t]) == 0},
      {C[1], C[2]}] /. t → 0 // Simplify // Flatten
```

$$\text{Out[133]= } \left\{ C[1] \to \frac{a\left(-1 + \frac{\beta}{\sqrt{\beta^2-\omega 0^2}}\right)}{2\,\omega 0^2},\; C[2] \to -\frac{a + \frac{a\beta}{\sqrt{\beta^2-\omega 0^2}}}{2\,\omega 0^2} \right\}$$

Substituting back C[1] and C[2] into the general solution, $x(t)$ becomes

```
In[134]:= solutionStepDriven = x[t] /. solutionStepDriven /.
    {C[1] → (C[1] /. constants[[1]]), C[2] → (C[2] /. constants[[2]])} // Simplify
```

$$\text{Out[134]= } \left\{ \frac{a + \frac{1}{2}e^{-t\left(\beta+\sqrt{\beta^2-\omega 0^2}\right)}\left(a\left(-1+\frac{\beta}{\sqrt{\beta^2-\omega 0^2}}\right) - e^{2t\sqrt{\beta^2-\omega 0^2}}\left(a + \frac{a\beta}{\sqrt{\beta^2-\omega 0^2}}\right)\right)}{\omega 0^2} \right\}$$

This response solution is shown in **Figure 2.3.11** below for $\beta = 0.2$ and $\beta = 0$ (no damping presence)

```
In[135]:= Plot[Evaluate[Table[solutionStepDriven /. {a → 2, ω0 → 1, β → i}, {i, {0.2, 0}}]],
    {t, 0, 30}, PlotRange → Full,
    PlotStyle → {{Blue, Thickness[0.005]}, {Dotted, Black, Thickness[0.005]}},
    Epilog → {Text["β=0.2", {3, 1.4}], Text["β=0", {12, 3}]}]
```

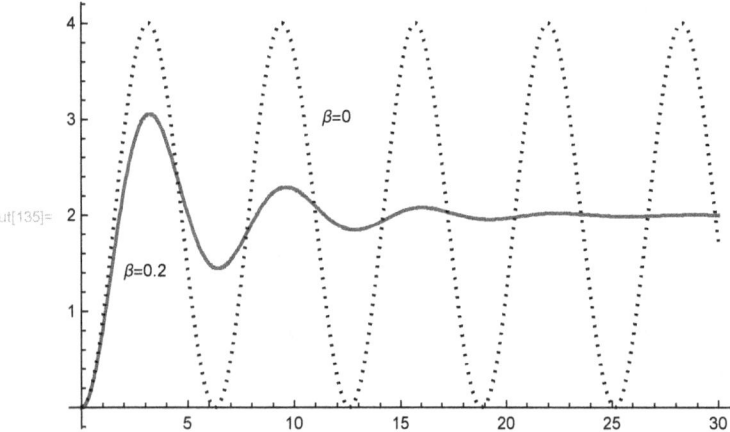

Figure 2.3.11. Response function solution to the step force function, for $\beta = 0.2$ and $\beta = 0$

The transient-state condition is represented by the complementary solution $x_c(t)$, is often of considerable importance for many types of physical applications.

At large times, the steady-state condition of the oscillator is represented by the particular solution $x_p(t)$, which is simply a displacement by an amount of a/ω_0^2.

If no damping occurs, that is $\beta = 0$, the oscillation is simply sinusoidal.

2.3.4.2 Response to an Impulsive Force Function

Suppose now the driving force is an impulsive function that has the width $t_1 - t_0 = \tau$, then, because the differential equation governing the motion of the oscillator is linear, the general solution for $t > t_1$ is given by the superposition of the solution for the positive step function applied at $t = t_0$, and the solution for the negative step function applied at $t = t_1$.

Let us consider $t_0 = 0$, so that $t_1 = \tau$, and the solution of the differential equation for $t > t_1$ is

```
In[136]:= solutionImpulseDriven =
    solutionStepDriven - (solutionStepDriven /. t → (t - τ)) // Simplify
```

$$\text{Out[136]}= \left\{ \frac{1}{2\,\omega 0^2} \left(e^{-t\left(\beta+\sqrt{\beta^2-\omega 0^2}\right)} \left(a\left(-1+\frac{\beta}{\sqrt{\beta^2-\omega 0^2}}\right) - e^{2t\sqrt{\beta^2-\omega 0^2}} \left(a+\frac{a\,\beta}{\sqrt{\beta^2-\omega 0^2}}\right)\right) + e^{-(t-\tau)\left(\beta+\sqrt{\beta^2-\omega 0^2}\right)} \left(a - \frac{a\,\beta}{\sqrt{\beta^2-\omega 0^2}} + e^{2(t-\tau)\sqrt{\beta^2-\omega 0^2}}\left(a+\frac{a\,\beta}{\sqrt{\beta^2-\omega 0^2}}\right)\right)\right)\right\}$$

Let us now plot the response function solution to the impulse force function as follows

```
In[137]:= Plot[If[0 < t < τ, solutionStepDriven, solutionImpulseDriven] /.
    {τ → 30, a → 2, ω0 → 1, β → 0.2}, {t, 0, 60},
    AxesLabel → {"t", "x(t)"}, PlotRange → Full]
```

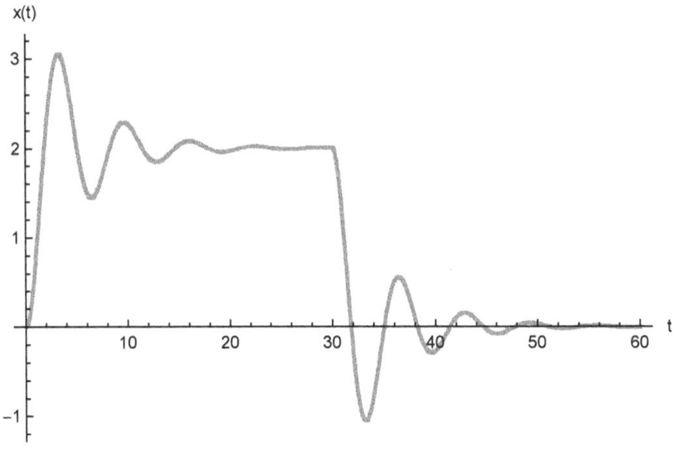

Figure 2.3.12. Response function solution to the impulse force function ($a = 2$, $\omega_0 = 1$, $\beta = 0.2$)

Notice from Figure 2.3.12, the oscillator's response to an impulse driving force is about 30 seconds. This duration corresponds to the width of the impulse force function, τ. However, as t becomes large, the oscillator returns to the state of equilibrium.

This page intentionally left blank

Chapter 3. Nonlinear Oscillatory Systems

Initialization

In[1]:= `Clear["Global`*"]`

3.1 Simple Nonlinear Oscillator

Consider a simple oscillator system such as the one we have studied in section 2.1. Let a deviation of the force from linearity be present in the system such that

(3.1.1) $\qquad F(x) = -kx + \epsilon x^3$,

where k and ϵ are constants; ϵ is usually a small quantity.

In[2]:= `(* Equation 3.1.1 *)`
`F[x_] := -k x + ϵ x^3`

The potential corresponding to such nonlinear force is

In[3]:= $V = -\int_0^x F[x]\, dx$

Out[3]= $\dfrac{k x^2}{2} - \dfrac{x^4 \epsilon}{4}$

Note that the potential is symmetric. Depending on the sign of ϵ, the force may either be greater or less than the linear approximation.

Let us plot the graph of comparison between the two potentials

Chapter 3. Nonlinear Oscillatory Systems

In[4]:= `Plot[Evaluate[{V, 1/2 k x^2} /. {k → 1, ϵ → 0.2}], {x, -6, 6},`
   `PlotStyle → {Blue, {Red, Dashed}}, AxesLabel → {"x", "V"}]`

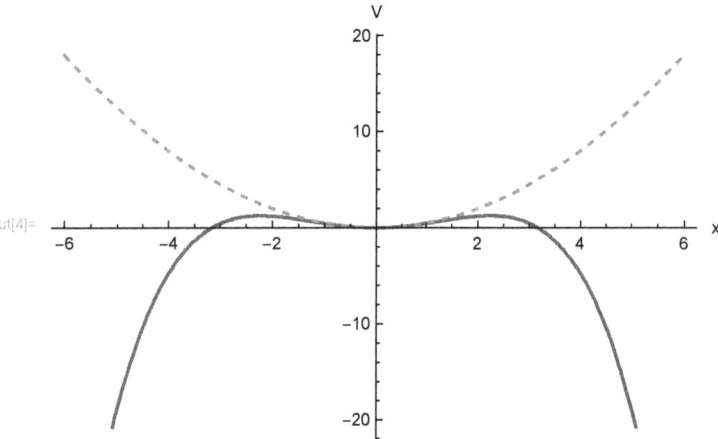

Out[4]=

Figure 3.1.1. The graph of the potential $V = \frac{1}{2}kx^2 - \frac{1}{4}\epsilon x^4$ (blue) and $\frac{1}{2}kx^2$ (red), indicating a parabolic region where simple harmonic motion is applicable, that is, the region for small displacement

The differential equation that governs the motion becomes

$$\frac{d^2 x}{dt^2} = -\omega_0^2 x + \alpha x^3,$$

where $\omega_0^2 = \frac{k}{m}$ and $\alpha = \frac{\epsilon}{m}$.

In[5]:= `eq31 = D[x[t], {t, 2}] == -ω0^2 x[t] + α x[t]^3`

Out[5]= $x''[t] == -\omega 0^2 x[t] + \alpha x[t]^3$

In[6]:= `(* Solve the differential equation *)`
`sol31 = DSolve[eq31, x, t] // PowerExpand`

    Solve: Inverse functions are being used by Solve, so some solutions may not be found; use Reduce for complete solution information.

Out[6]= $\left\{\left\{x \to \text{Function}\left[\{t\}, -\dfrac{1}{\omega 0^2 + \sqrt{\omega 0^4 - 2\alpha C[1]}} 2 i C[1] \left(\sqrt{-1} \left(\sqrt{\alpha} \left(\omega 0^2 - \sqrt{\omega 0^4 - 2\alpha C[1]}\right)\right)^{-\frac{1}{2}}\right) \text{JacobiSN}\left[\dfrac{1}{\sqrt{2}} \left(\sqrt{\left(t^2 \omega 0^2 + t^2 \sqrt{\omega 0^4 - 2\alpha C[1]} + 2 t \omega 0^2 C[2] + 2 t \sqrt{\omega 0^4 - 2\alpha C[1]}\, C[2] + \omega 0^2 C[2]^2 + \sqrt{\omega 0^4 - 2\alpha C[1]}\, C[2]^2}\right)\right), \dfrac{\omega 0^2 - \sqrt{\omega 0^4 - 2\alpha C[1]}}{\omega 0^2 + \sqrt{\omega 0^4 - 2\alpha C[1]}}\right]\right]\right\}, $
$\left\{x \to \text{Function}\left[\{t\}, \dfrac{1}{\omega 0^2 + \sqrt{\omega 0^4 - 2\alpha C[1]}} 2 i C[1] \left(\sqrt{-1} \left(\sqrt{\alpha}\left(\omega 0^2 - \sqrt{\omega 0^4 - 2\alpha C[1]}\right)\right)^{-\frac{1}{2}}\right) \text{JacobiSN}\left[\dfrac{1}{\sqrt{2}} \left(\sqrt{\left(t^2 \omega 0^2 + t^2 \sqrt{\omega 0^4 - 2\alpha C[1]} + 2 t \omega 0^2 C[2] + 2 t \sqrt{\omega 0^4 - 2\alpha C[1]}\, C[2] + \omega 0^2 C[2]^2 + \sqrt{\omega 0^4 - 2\alpha C[1]}\, C[2]^2}\right)\right), \dfrac{\omega 0^2 - \sqrt{\omega 0^4 - 2\alpha C[1]}}{\omega 0^2 + \sqrt{\omega 0^4 - 2\alpha C[1]}}\right]\right]\right\}\right\}$

Notice here that the solutions involve **Jacobi elliptic functions**.

In[7]:= `? JacobiSN`

    JacobiSN[u, m] gives the Jacobi elliptic function sn(u | m). ≫

We shall choose the first solution since it is positive. The second solution is the negative of the first one. The initial displacement can be found at $t = 0$ as follows

In[8]:= `xInit = x[0] /. sol31[[1]]`

Out[8]= $\dfrac{2\sqrt{\alpha}\, C[1]\, \text{JacobiSN}\left[\dfrac{\sqrt{\omega 0^2 C[2]^2 + \sqrt{\omega 0^4 - 2\alpha C[1]}\, C[2]^2}}{\sqrt{2}}, \dfrac{\omega 0^2 - \sqrt{\omega 0^4 - 2\alpha C[1]}}{\omega 0^2 + \sqrt{\omega 0^4 - 2\alpha C[1]}}\right]}{\sqrt{\omega 0^2 - \sqrt{\omega 0^4 - 2\alpha C[1]}}\, \left(\omega 0^2 + \sqrt{\omega 0^4 - 2\alpha C[1]}\right)}$

Chapter 3. Nonlinear Oscillatory Systems

In[9]:= `(* Plot the solution for a given conditions *)`
`Plot[x[t] /. sol31[[1]] /. {ω0 → 1, α → 0.2, C[1] → 1, C[2] → 1},`
`{t, 0, 20}, AxesLabel → {"t", "x"}, PlotRange → {{0, 20}, {-1.2, 1.2}}]`

Out[9]=

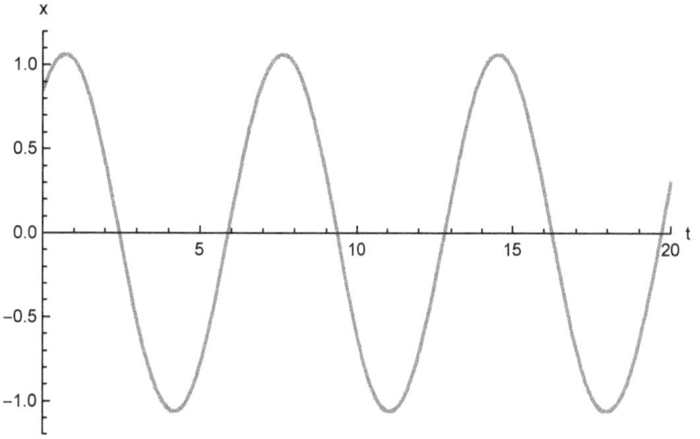

Figure 3.1.2. Plot the solution for a given set of conditions
($\omega_0 = 1$, $\alpha = 0.2$, $C[1] = 1$, $C[2] = 1$)

As for comparison, let us now plot the solutions for various amplitudes

In[10]:= `(* Plot the solutions for various amplitudes *)`
`Plot[Evaluate[Table[x[t] /. sol31[[1]] /. {ω0 → 1, α → 0.2, C[1] → i, C[2] → 1},`
`{i, 0.5, 4, 0.5}]], {t, 0, 4 π}, AxesLabel → {"t", "x"}]`

--- Plot: {(1. + 2 t + t²) - 0, Indeterminate - 0, Im[2 t + t²] - 0} must be a list of equalities or real-valued functions.

Out[10]=

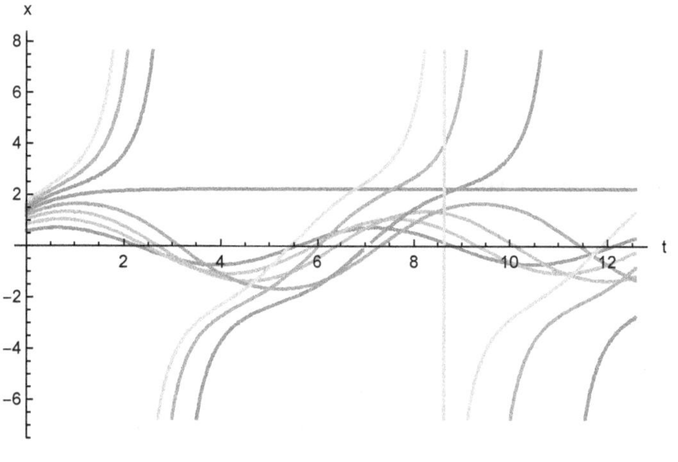

Figure 3.1.3. The graph of the solutions for various amplitudes ($\omega_0 = 1$, $\alpha = 0.2$, $C[2] = 1$)

We notice here in **Figure 3.1.3**, there are 3 types of motion possible; these are oscillations, asymptotic motion, and revolutions. The asymptotic motion separates the other two motions.

Note also that for oscillatory motion, it is periodic. However, the period of oscillation now depends on the amplitude. We shall see later that the period of the motion is a function of the amplitude.

The total energy of the system is $H = \frac{1}{2} m \left(\frac{dx}{dt}\right)^2 + V$.

$$dt = \sqrt{\frac{m}{2}} \, \frac{dx}{\sqrt{H-V}}.$$

At $t = 0$, let x_0 be the initial amplitude and let the initial velocity be zero so that the kinetic energy is zero at $t = 0$. Thus, $H = V$ at $x = \pm x_0$.

The period can be calculated from

$$\text{period} = \int_0^T dt = 4 \sqrt{\frac{m}{2}} \int_0^{x_0} \frac{1}{\sqrt{H-V}} \, dx,$$

where the total energy H equals to V at $t = 0$,

In[11]:= `(* Total energy equals to V at t = 0 *)`
`V /. x -> x0`

Out[11]= $\dfrac{k\, x0^2}{2} - \dfrac{x0^4\, \epsilon}{4}$

We define **time(x)** as a function of displacement and substitute H, the total energy of the system

In[12]:= `(* Time required to move the particle from one point to another *)`

$\text{time}[x\_] = \sqrt{\dfrac{m}{2}} \int \dfrac{1}{\sqrt{H-V}} \, dx \;/.\; H \to (V \;/.\; x \to x0) \; // \; \text{Simplify} \; // \; \text{PowerExpand}$

Out[12]= $-\dfrac{i\sqrt{2}\,\sqrt{m}\,\sqrt{2k - x^2\epsilon - x0^2\epsilon}\,\sqrt{x^2\epsilon - x0^2\epsilon}\, \text{EllipticF}\!\left[\text{ArcSin}\!\left[\dfrac{x}{x0}\right],\, \dfrac{x0^2\epsilon}{2k - x0^2\epsilon}\right]}{\sqrt{x^2 - x0^2}\,\sqrt{\epsilon}\,\sqrt{2k - x0^2\epsilon}\,\sqrt{-2k + (x^2 + x0^2)\epsilon}}$

Notice here that the result of the integration involves a special function **EllipticF**

In[13]:= `? EllipticF`

EllipticF[ϕ, m] gives the elliptic integral of the first kind $F(\phi \mid m)$. »

Since the motion is symmetric, the period of oscillation is 4 times the integration over x from $x = x_0$ to $x = 0$.

However, since the value of the function **time(x)** maybe undefined at x_0, we take the limit of the integration as it approaches x_0. Hence

Chapter 3. Nonlinear Oscillatory Systems

$$\int_0^T dt = 4 \lim_{x \to x_0} \sqrt{\frac{m}{2}} \int_0^x \frac{1}{\sqrt{H-V}} dx$$

In[14]:= `(* Period as a function of initial amplitude, x0 *)`
`period[x0_] =`
`4 Limit[time[0] - time[x], x → x0] /. {k → ω0² m, ϵ → α m} // Simplify // PowerExpand`

Out[14]= $\dfrac{4 \, i \, \sqrt{2} \, \sqrt{-x0^2 \alpha + \omega 0^2} \; \text{EllipticK}\!\left[-\dfrac{x0^2 \alpha}{x0^2 \alpha - 2 \, \omega 0^2}\right]}{\sqrt{x0^2 \alpha - \omega 0^2} \, \sqrt{-x0^2 \alpha + 2 \, \omega 0^2}}$

The following graph shows the period as a function of the amplitude

In[15]:= `Plot[period[x0] /. {ω0 → 1, α → 0.2} // Chop, {x0, 0, 2},`
`  AxesLabel → {"x₀", "Period"}, PlotRange → {{0, 2}, {0, 3 π}}]`

Out[15]=

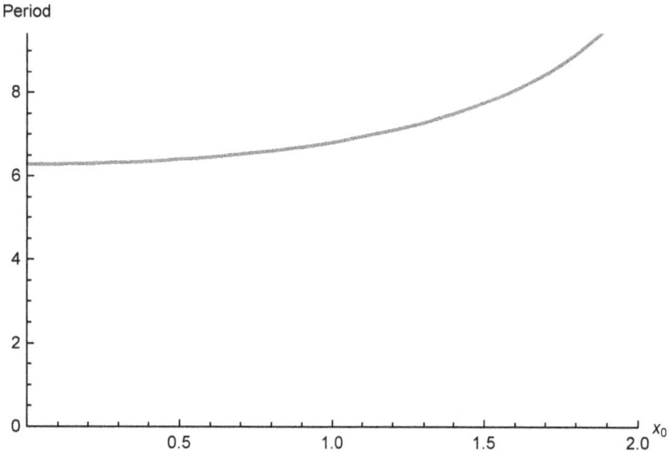

Figure 3.1.4. The graph shows the relation between the period of motion with the amplitude of a nonlinear oscillator.

3.1.1 Representation on the Phase Plane

In[16]:= `(* Define the phase points on the phase plane x vs ẋ *)`
`phasePoints = Flatten[{x[t], D[x[t], t]} /. sol31] // Simplify // PowerExpand`

Out[16]= $\left\{ \left(2\sqrt{\alpha}\, C[1]\, \text{JacobiSN}\left[\frac{\sqrt{\omega 0^2 + \sqrt{\omega 0^4 - 2\alpha C[1]}}\, (t+C[2])}{\sqrt{2}},\, \frac{\omega 0^2 - \sqrt{\omega 0^4 - 2\alpha C[1]}}{\omega 0^2 + \sqrt{\omega 0^4 - 2\alpha C[1]}} \right] \right) \right/$

$\left(\sqrt{\omega 0^2 - \sqrt{\omega 0^4 - 2\alpha C[1]}}\, \left(\omega 0^2 + \sqrt{\omega 0^4 - 2\alpha C[1]}\right) \right),$

$\left(\sqrt{2}\sqrt{\alpha}\, C[1]\, \text{JacobiCN}\left[\frac{\sqrt{\omega 0^2 + \sqrt{\omega 0^4 - 2\alpha C[1]}}\, (t+C[2])}{\sqrt{2}},\, \frac{\omega 0^2 - \sqrt{\omega 0^4 - 2\alpha C[1]}}{\omega 0^2 + \sqrt{\omega 0^4 - 2\alpha C[1]}} \right] \right.$

$\left. \text{JacobiDN}\left[\frac{\sqrt{\omega 0^2 + \sqrt{\omega 0^4 - 2\alpha C[1]}}\, (t+C[2])}{\sqrt{2}},\, \frac{\omega 0^2 - \sqrt{\omega 0^4 - 2\alpha C[1]}}{\omega 0^2 + \sqrt{\omega 0^4 - 2\alpha C[1]}} \right] \right) \Big/$

$\left(\sqrt{\omega 0^2 - \sqrt{\omega 0^4 - 2\alpha C[1]}}\, \sqrt{\omega 0^2 + \sqrt{\omega 0^4 - 2\alpha C[1]}} \right),$

$-\left(\left(2\sqrt{\alpha}\, C[1]\, \text{JacobiSN}\left[\frac{1}{\sqrt{2}}\sqrt{\omega 0^2 + \sqrt{\omega 0^4 - 2\alpha C[1]}}\, (t+C[2]),\, \frac{\omega 0^2 - \sqrt{\omega 0^4 - 2\alpha C[1]}}{\omega 0^2 + \sqrt{\omega 0^4 - 2\alpha C[1]}} \right] \right) \right/$

$\left(\sqrt{\omega 0^2 - \sqrt{\omega 0^4 - 2\alpha C[1]}}\, \left(\omega 0^2 + \sqrt{\omega 0^4 - 2\alpha C[1]}\right) \right) \right),$

$-\left(\left(\sqrt{2}\sqrt{\alpha}\, C[1]\, \text{JacobiCN}\left[\frac{1}{\sqrt{2}}\sqrt{\omega 0^2 + \sqrt{\omega 0^4 - 2\alpha C[1]}}\, (t+C[2]),\, \frac{\omega 0^2 - \sqrt{\omega 0^4 - 2\alpha C[1]}}{\omega 0^2 + \sqrt{\omega 0^4 - 2\alpha C[1]}} \right] \right.\right.$

$\left. \text{JacobiDN}\left[\frac{1}{\sqrt{2}}\sqrt{\omega 0^2 + \sqrt{\omega 0^4 - 2\alpha C[1]}}\, (t+C[2]),\, \frac{\omega 0^2 - \sqrt{\omega 0^4 - 2\alpha C[1]}}{\omega 0^2 + \sqrt{\omega 0^4 - 2\alpha C[1]}} \right] \right) \Big/$

$\left. \left(\sqrt{\omega 0^2 - \sqrt{\omega 0^4 - 2\alpha C[1]}}\, \sqrt{\omega 0^2 + \sqrt{\omega 0^4 - 2\alpha C[1]}} \right) \right) \Big\}$

Chapter 3. Nonlinear Oscillatory Systems

```
In[17]:= (* Phase plane plot for various initial conditions *)
ParametricPlot[
  Evaluate[Table[phasePoints /. {ω0 → 1, α → 0.2, C[1] → i, C[2] → 1}, {i, 0.5, 4, 0.5}]],
  {t, -2 π, 2 π}, AxesLabel → {"x ", "x˙"}, PlotRange → {{-4, 4}, {-3, 3}}]
```

--- ParametricPlot: {Indeterminate - 0} must be a list of equalities or real-valued functions.

--- ParametricPlot: {Indeterminate - 0} must be a list of equalities or real-valued functions.

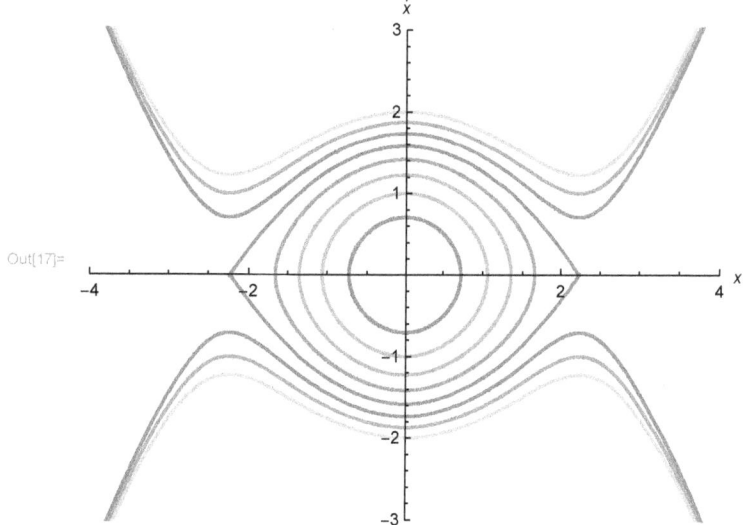

Figure 3.1.5. Phase space diagram of a simple nonlinear oscillator for various amplitudes ($\omega_0 = 1$, $\alpha = 0.2$, $C[2] = 1$)

The phase space diagram shows that near the center, there exist oscillations such that the motions are bounded. Where as for larger energies, the motions are unbounded (asymptotic behavior). In between them, there is a curve called **separatrix** that separates bounded and unbounded motions.

3.2 Simple Nonlinear Pendulum

Another example of a simple nonlinear system is nonlinear pendulum. Suppose a particle of mass m is connected by a massless string to a rigid support. Let ϕ be the angle that the string makes with the vertical and assume that the there are only two forces acting on the particle, gravity and the tension of the string.

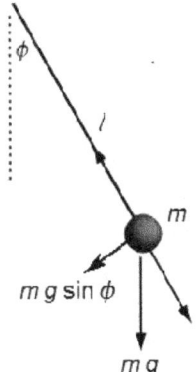

The force perpendicular to the string is given by

$$F_\phi = -mg \sin \phi,$$

where g is the acceleration due to gravity.

According to Newton's second law, the motion is described by

$$m \frac{d^2 s}{dt^2} = -mg \sin \phi,$$

where s is the displacement in arc-length.

Since the displacement along the arc is $s = l\phi$, where l is the length of the string, the equation of motion becomes

(3.2) $$\frac{d^2 \phi}{dt^2} = -\frac{g}{l} \sin \phi.$$

If the displacement is small so that the angle that the string makes with the vertical is also small, we have the approximation $\sin \phi \approx \phi$. Thus, the solution is reduced to simple harmonic motion. Our analysis in Chapter 2 for simple harmonic oscillator is applicable for such case in which the natural angular frequency becomes $\omega_0 = \sqrt{g/l}$. Within this approximation, the oscillation period T is

$$T = 2\pi/\omega_0 = 2\pi \sqrt{l/g},$$

and it is independent of the amplitude. This is the key feature of simple harmonic oscillation.

However, in many physical situations, the mass is not required to oscillate in small angles.

Thus, we now consider a simple nonlinear pendulum where it can swing to large angles. For our present case we omit damping and consider system without driving force. Since there is no adding or removing of energy from this system, the total energy is conserved and the pendulum executes a periodic motion. Its period, however, is no longer independent of amplitude.

The potential energy due to gravity for this system is

$$V = mgl(1 - \cos \phi)$$

```
In[18]:= Clear[ϕ, ω, time, period]
```

Chapter 3. Nonlinear Oscillatory Systems

In[19]:= **V = m g l (1 - Cos[ϕ])**

Out[19]= $g\, l\, m\, (1 - \cos[\phi])$

The equation of motion becomes

In[20]:= **(* Equation 3.2 *)**
eq32 = ϕ''[t] + ω0² Sin[ϕ[t]] == 0

Out[20]= $\omega 0^2 \sin[\phi[t]] + \phi''[t] == 0$

In[21]:= **(* The solution *)**
sol32 = DSolve[ϕ''[t] + ω0² Sin[ϕ[t]] == 0, ϕ, t]

··· Solve: Inverse functions are being used by Solve, so some solutions may not be found; use Reduce for complete solution information.

Out[21]= $\left\{\left\{\phi \to \text{Function}\left[\{t\}, -2\, \text{JacobiAmplitude}\left[\frac{1}{2}\sqrt{(2\,\omega 0^2 + C[1])\,(t+C[2])^2},\, \frac{4\,\omega 0^2}{2\,\omega 0^2 + C[1]}\right]\right]\right\},$
$\left\{\phi \to \text{Function}\left[\{t\}, 2\, \text{JacobiAmplitude}\left[\frac{1}{2}\sqrt{(2\,\omega 0^2 + C[1])\,(t+C[2])^2},\, \frac{4\,\omega 0^2}{2\,\omega 0^2 + C[1]}\right]\right]\right\}\right\}$

In[22]:= **? JacobiAmplitude**

JacobiAmplitude[*u*, *m*] gives the amplitude am(*u* | *m*) for Jacobi elliptic functions. ≫

We notice that the solutions involve the amplitudes for Jacobi elliptic functions. We shall choose the positive solution

In[23]:= **ϕ[t] /. sol32[[2]]**

Out[23]= $2\, \text{JacobiAmplitude}\left[\frac{1}{2}\sqrt{(2\,\omega 0^2 + C[1])\,(t+C[2])^2},\, \frac{4\,\omega 0^2}{2\,\omega 0^2 + C[1]}\right]$

The initial amplitude can be found at *t* = 0

In[24]:= **ϕInit = ϕ[0] /. sol32[[2]]**

Out[24]= $2\, \text{JacobiAmplitude}\left[\frac{1}{2}\sqrt{(2\,\omega 0^2 + C[1])\,C[2]^2},\, \frac{4\,\omega 0^2}{2\,\omega 0^2 + C[1]}\right]$

The maximum value of the solution *ϕ*(*t*) can be found by using **NMaxValue** command

Chapter 3. Nonlinear Oscillatory Systems

In[25]:= **NMaxValue[ϕ[t] /. sol32[[2]], t] /. {ω0 → 1, C[1] → 1, C[2] → 1}**

--- NMaxValue: The function value $-2\,\text{JacobiAmplitude}\left[\frac{1}{2}\sqrt{(2\omega0^2+C[1])(-0.829053+C[2])^2}, \frac{4\omega0^2}{2\omega0^2+C[1]}\right]$ is not a number at $\{t\} = \{-0.829053\}$.

--- NMaxValue: The function value $-2\,\text{JacobiAmplitude}\left[\frac{1}{2}\sqrt{(2\omega0^2+C[1])(-0.829053+C[2])^2}, \frac{4\omega0^2}{2\omega0^2+C[1]}\right]$ is not a number at $\{t\} = \{-0.829053\}$.

--- NMaxValue: The function value $-2\,\text{JacobiAmplitude}\left[\frac{1}{2}\sqrt{(2\omega0^2+C[1])(-0.829053+C[2])^2}, \frac{4\omega0^2}{2\omega0^2+C[1]}\right]$ is not a number at $\{t\} = \{-0.829053\}$.

--- General: Further output of NMaxValue::nnum will be suppressed during this calculation.

Out[25]= 2.0944

We can use **NArgMax** command to find *t* at which ϕ(*t*) is maximum.

In[26]:= **NArgMax[ϕ[t] /. sol32[[2]], t] /. {ω0 → 1, C[1] → 1, C[2] → 1} // Chop**

--- NArgMax: The function value $-2\,\text{JacobiAmplitude}\left[\frac{1}{2}\sqrt{(2\omega0^2+C[1])(-0.829053+C[2])^2}, \frac{4\omega0^2}{2\omega0^2+C[1]}\right]$ is not a number at $\{t\} = \{-0.829053\}$.

--- NArgMax: The function value $-2\,\text{JacobiAmplitude}\left[\frac{1}{2}\sqrt{(2\omega0^2+C[1])(-0.829053+C[2])^2}, \frac{4\omega0^2}{2\omega0^2+C[1]}\right]$ is not a number at $\{t\} = \{-0.829053\}$.

--- NArgMax: The function value $-2\,\text{JacobiAmplitude}\left[\frac{1}{2}\sqrt{(2\omega0^2+C[1])(-0.829053+C[2])^2}, \frac{4\omega0^2}{2\omega0^2+C[1]}\right]$ is not a number at $\{t\} = \{-0.829053\}$.

--- General: Further output of NArgMax::nnum will be suppressed during this calculation.

Out[26]= 1.15652

In[27]:= `(* Plot the solution for the given conditions *)`
`Plot[ϕ[t] /. sol32[[2]] /. {ω0 → 1, C[1] → 1, C[2] → 1},`
`{t, 0, 4 π}, AxesLabel → {"t", "ϕ"}]`

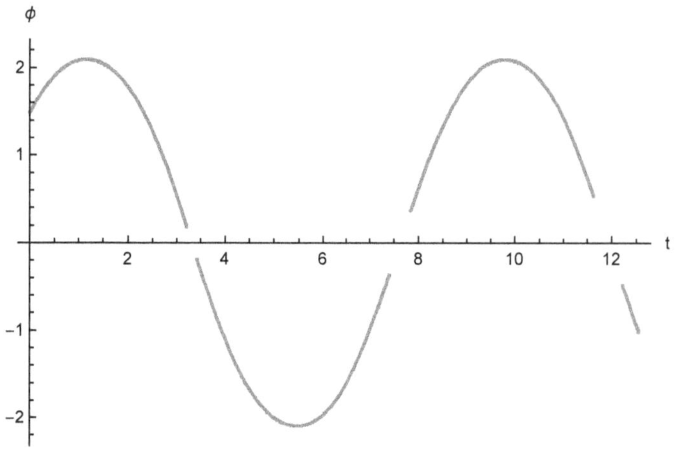

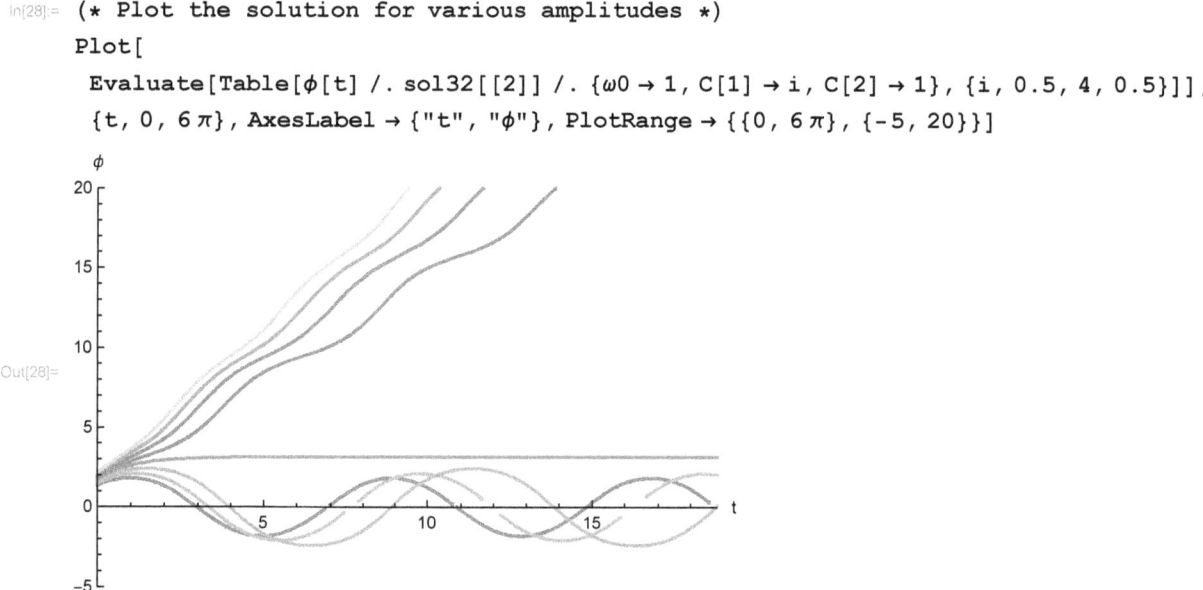

Figure 3.2.1. The graph of the solution of a simple nonlinear pendulum for the given set of conditions ($\omega_0 = 1$, $C[1] = 1$, $C[2] = 1$)

In[28]:= `(* Plot the solution for various amplitudes *)`
`Plot[`
`Evaluate[Table[ϕ[t] /. sol32[[2]] /. {ω0 → 1, C[1] → i, C[2] → 1}, {i, 0.5, 4, 0.5}]],`
`{t, 0, 6 π}, AxesLabel → {"t", "ϕ"}, PlotRange → {{0, 6 π}, {-5, 20}}]`

Figure 3.2.2. The graph of the solutions for various amplitudes ($\omega_0 = 1$, $C[2] = 1$)

We notice from Figure 3.2.2, there are three types of motions just as the case with nonlinear oscillator; these are: oscillations, asymptotic, and revolutions. The asymptotic one is between oscillations and revolutions. For oscillatory motions, the period is not constant, but depends on the amplitudes.

We shall apply the same method as in Section 3.1 to find the period of oscillation

In[29]:= `(* Time required to move the particle from one point to another *)`
`time[ϕ_] =` $\sqrt{\dfrac{m}{2}} \int \dfrac{1}{\sqrt{H-V}}\, d\phi$ `/. H → (V /. ϕ → ϕ0) // Simplify // PowerExpand`

Out[29]= $\dfrac{\sqrt{2}\,\sqrt{-\cos[\phi]+\cos[\phi 0]}\;\text{EllipticF}\!\left[\dfrac{\phi}{2},\,\csc\!\left[\dfrac{\phi 0}{2}\right]^{2}\right]}{\sqrt{g}\,\sqrt{l}\,\sqrt{\cos[\phi]-\cos[\phi 0]}\,\sqrt{-1+\cos[\phi 0]}}$

In[30]:= `(* Period as a function of intial amplitude, ϕ0 *)`
`period[ϕ0_] = 4 Limit[time[0] - time[ϕ], ϕ → ϕ0] // Simplify // PowerExpand`

Out[30]= $\dfrac{4\,i\,\sqrt{2}\;\text{EllipticF}\!\left[\dfrac{\phi 0}{2},\,\csc\!\left[\dfrac{\phi 0}{2}\right]^{2}\right]}{\sqrt{g}\,\sqrt{l}\,\sqrt{-1+\cos[\phi 0]}}$

In[31]:= `Plot[period[ϕ0] /. {g → 9.8, l → 1}, {ϕ0, 0, 2},`
`  AxesLabel → {"ϕ0", "Period"}, PlotRange → {{0, 2}, {0, 3}}, PlotPoints → 10]`

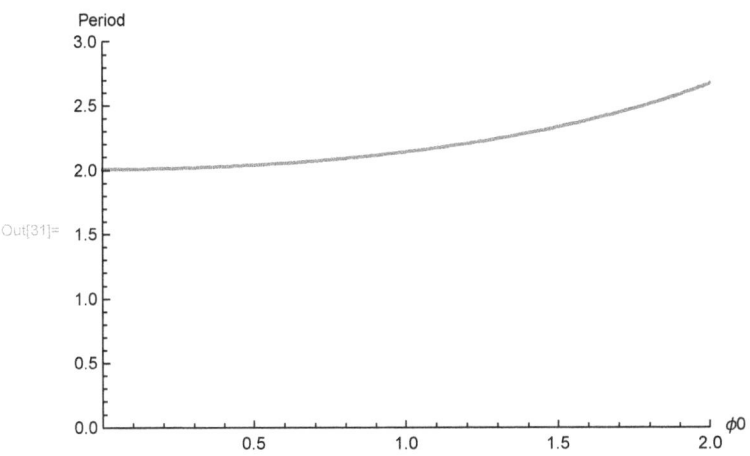

Figure 3.2.3. The graph shows the relation between the period of motion with the amplitude of a nonlinear pendulum.

3.2.1 Representation on the Phase Plane

Chapter 3. Nonlinear Oscillatory Systems

```
In[32]:= (* Define the phase points *)
        phasePoints = Flatten[{ϕ[t], D[ϕ[t], t]} /. sol32] // Simplify // PowerExpand
```

$$\text{Out[32]= } \left\{ -2 \text{ JacobiAmplitude}\left[\frac{1}{2}\sqrt{2\,\omega 0^2 + C[1]}\ (t + C[2]),\ \frac{4\,\omega 0^2}{2\,\omega 0^2 + C[1]}\right], \right.$$

$$-\sqrt{2\,\omega 0^2 + C[1]}\ \text{JacobiDN}\left[\frac{1}{2}\sqrt{2\,\omega 0^2 + C[1]}\ (t + C[2]),\ \frac{4\,\omega 0^2}{2\,\omega 0^2 + C[1]}\right],$$

$$2\ \text{JacobiAmplitude}\left[\frac{1}{2}\sqrt{2\,\omega 0^2 + C[1]}\ (t + C[2]),\ \frac{4\,\omega 0^2}{2\,\omega 0^2 + C[1]}\right],$$

$$\left. \sqrt{2\,\omega 0^2 + C[1]}\ \text{JacobiDN}\left[\frac{1}{2}\sqrt{2\,\omega 0^2 + C[1]}\ (t + C[2]),\ \frac{4\,\omega 0^2}{2\,\omega 0^2 + C[1]}\right]\right\}$$

```
In[33]:= (* Phase plane plot for various initial conditions
        *)
        ParametricPlot[
         Evaluate[Table[phasePoints /. {ω0 → 1, C[1] → i, C[2] → 1}, {i, 0.5, 4, 0.5}]],
         {t, -6 π, 6 π}, AxesLabel → {"ϕ", "ϕ'"}, PlotRange → {{-5, 5}, {-3, 3}}]
```

--- ParametricPlot: {Indeterminate − 0} must be a list of equalities or real-valued functions.

--- ParametricPlot: {Indeterminate − 0} must be a list of equalities or real-valued functions.

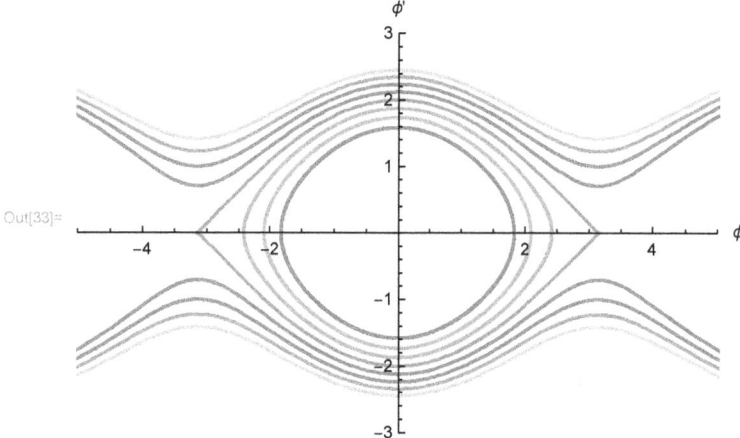

Figure 3.2.4. Phase space diagram for simple nonlinear pendulum. Notice the regions of bounded and unbounded motion. Between the bounded and unbounded regions is the **separatrix**.

3.3 Damped, Driven Nonlinear Pendulum

In this section, we shall consider more realistic pendulum when we put all three effects—damping, driving force, and the nonlinearity together at the same time. The situation becomes more complicated and also more interesting.

First, we do not assume small-angle approximation. Second, we include damping force of the form $-\gamma\,\dot{\phi}$. Third, we add to our model, a sinusoidal driving force $F_0\cos(\Omega\,t)$. Putting all these forces together, out equation of motion becomes,

(3.3.1) $$\frac{d^2\phi}{dt^2} = -\frac{g}{l}\sin\phi - \gamma\dot{\phi} + F_0\cos(\Omega t)$$

It can be separated into two first order differential equation

(3.3.2) $$\frac{d\omega}{dt} = -\frac{g}{l}\sin\phi - \gamma\omega + F_0\cos(\Omega t),$$

(3.3.3) $$\frac{d\phi}{dt} = \omega.$$

We shall solve this numerically, since there is no known exact solution to equation (3.3.1).

```
In[34]:= Clear[ϕ, ω]
```

The system of equations in Mathematica become

```
In[35]:= eq331 = D[ϕ[t], t] == ω[t]
Out[35]= ϕ'[t] == ω[t]
```

```
In[36]:= eq332 = D[ω[t], t] == -ω0^2 Sin[ϕ[t]] - γ ω[t] + F0 Cos[Ω t]
Out[36]= ω'[t] == F0 Cos[t Ω] - ω0^2 Sin[ϕ[t]] - γ ω[t]
```

where $\omega_0 = \sqrt{g/l}$. In which γ, F_0, and Ω are the damping factor, the amplitude of the driving force, and the driving frequency, respectively.

The solutions of the above system of equations are: $\phi(t)$ and $\omega(t)$, which are the coordinates of the phase space of our nonlinear motion.

3.3.1 Representation on the Phase Plane

To solve the equations numerically, we need to provide **NDSolve** command specific values of the parameters.

Let us consider the following cases based upon the parameters being used. In all cases we consider $g = l$, so that $\omega_0 = 1$.

Case (a). $F_0 = 0$.

In this case, there is no driving force.
Consider $\gamma = 1/2$, $\Omega = 2/3$

```
In[37]:= phasePoints1 =
    NDSolve[{eq331, eq332, ϕ[0] == 0.2, ω[0] == 0} /. {ω0 → 1, γ → 1/2, F0 → 0, Ω → 2/3},
     {ϕ, ω}, {t, 0, 90 π}] // Flatten
```

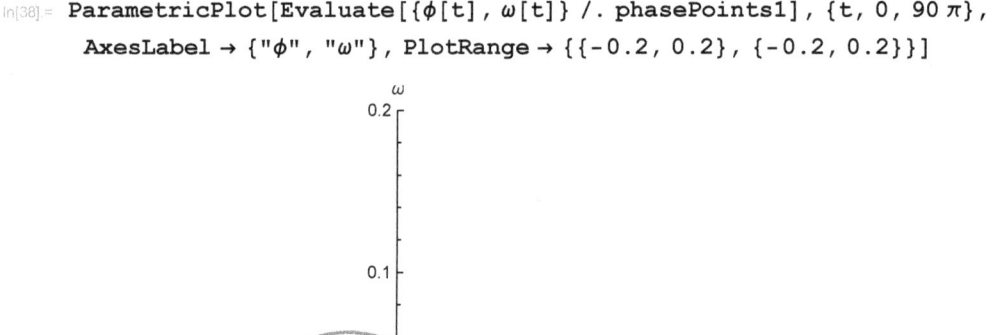

Out[37]= {ϕ → InterpolatingFunction[Domain: {{0., 283.}} Output: scalar],

ω → InterpolatingFunction[Domain: {{0., 283.}} Output: scalar]}

The phase diagram for the nonlinear pendulum when there is no driving force

```
In[38]:= ParametricPlot[Evaluate[{ϕ[t], ω[t]} /. phasePoints1], {t, 0, 90 π},
     AxesLabel → {"ϕ", "ω"}, PlotRange → {{-0.2, 0.2}, {-0.2, 0.2}}]
```

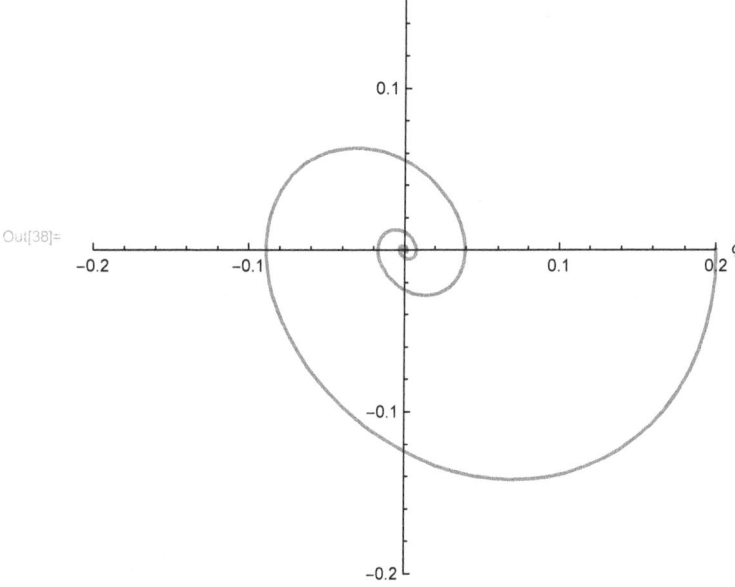

Figure 3.3.1. The phase diagram of a damped, non-driven nonlinear pendulum ($F_0 = 0$, $γ = 1/2$, $Ω = 2/3$)

For small damping factor, the phase diagram looks similar with the one for linear underdamped motion.

```
In[39]:= (* Plot the angle vs time, and
    the angular velocity vs time
*)
Plot[{ϕ[t] /. phasePoints1[[1]], ω[t] /. phasePoints1[[2]]} // Evaluate,
 {t, 0, 60}, PlotRange → Full]
```

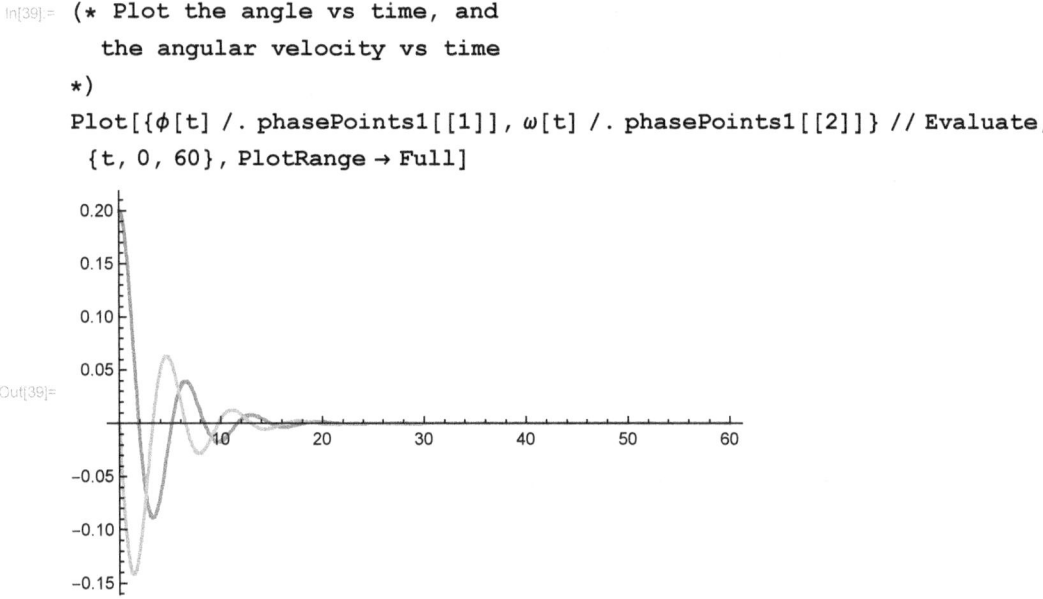

Figure 3.3.2. The graph of ϕ and ω as a function of time for a damped, non-driven nonlinear pendulum ($F_0 = 0$, $γ = 1/2$, $Ω = 2/3$)

Without a driving force, the motion is damped and the pendulum comes to rest after a few oscillations.

Case (b). $F_0 = 0.5$.

Consider $γ = 1/2$, $Ω = 2/3$

```
In[40]:= phasePoints2 =
  NDSolve[{eq331, eq332, ϕ[0] == 0.2, ω[0] == 0} /. {ω0 → 1, γ → 1/2, F0 → 0.5, Ω → 2/3},
   {ϕ, ω}, {t, 0, 90 π}] // Flatten
```

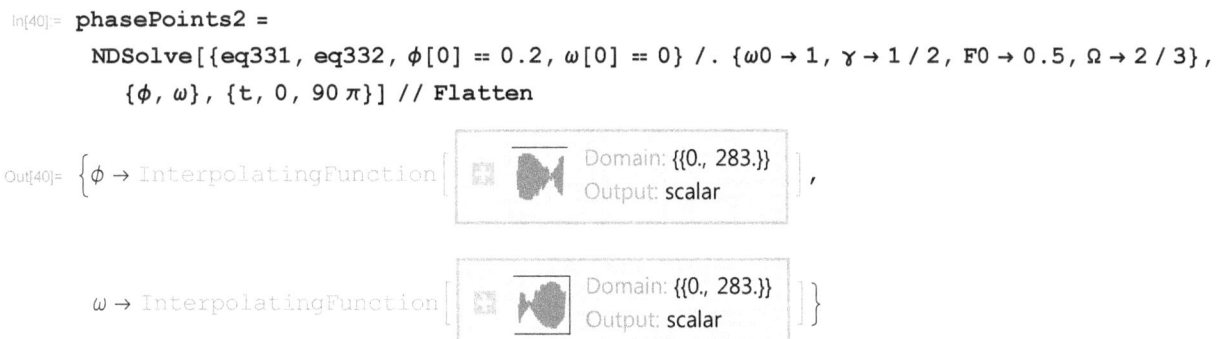

The phase diagram for the nonlinear pendulum when there is small driving force, $F_0 = 0.5$

In[41]:= `ParametricPlot[Evaluate[{ϕ[t], ω[t]} /. phasePoints2],`
`    {t, 0, 90 π}, AxesLabel → {"ϕ", "ω"}, PlotStyle → Blue, AspectRatio → 0.5]`

Out[41]=
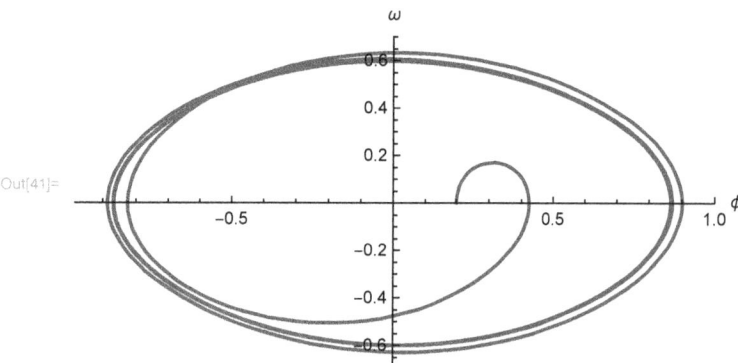

Figure 3.3.3. The phase diagram of a damped, driven nonlinear pendulum
($F_0 = 0.5$, $γ = 1/2$, $Ω = 2/3$)

The origin of the curve corresponds to the initial transient which depends on the initial conditions, where in this case we started with $ϕ(0) = 0.2$ and $ω(0) = 0$. However, the pendulum quickly settles into a regular orbit in phase space corresponding to the oscillatory motion of both $ϕ$ and $ω$. With some initial conditions we can solve for the result of the motion and it can be shown that the final orbit is independent of the initial conditions.

In[42]:= `(* Plot the angle vs time, and the angular velocity vs time *)`
`Plot[{ϕ[t] /. phasePoints2[[1]], ω[t] /. phasePoints2[[2]]} // Evaluate, {t, 0, 30 π},`
`    PlotRange → Full, PlotStyle → {Automatic, {Red, Dashed}}, AxesLabel → {"ϕ", "ω"}]`

Out[42]=
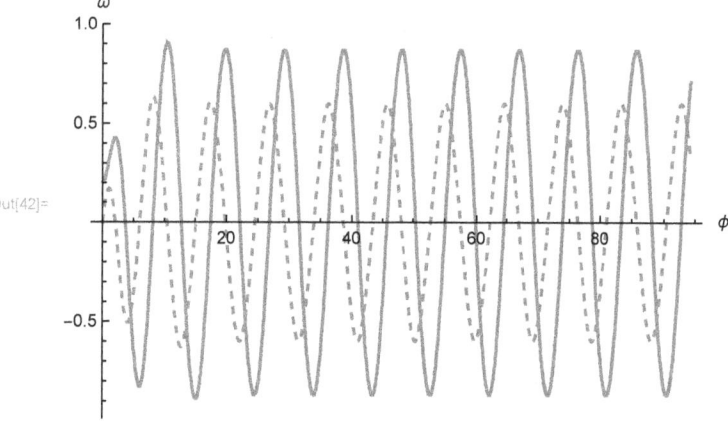

Figure 3.3.4. The graph of $ϕ$ and $ω$ as a function of time for a damped, driven nonlinear pendulum
($F_0 = 0.5$, $γ = 1/2$, $Ω = 2/3$)

With a small driving force, $F_0 = 0.5$, we find two regimes. The first regime is a first few oscillations affected by the damping force. It is initial transient as in the case of no driving force. The initial transient decays with time and has an angular frequency of approximately $ω_0$. After the transient is damped away, the pendulum settles into a steady oscillation in response to the driving force and then moves at the driving frequency Ω. This is very similar to the damped, driven, but linear oscillator discussed in Section 2.2.

Case (c). $F_0 = 1.2$

Consider $\gamma = 1/2$, $\Omega = 2/3$

```
In[43]:= phasePoints3 =
    NDSolve[{eq331, eq332, ϕ[0] == 0.2, ω[0] == 0} /. {ω0 → 1, γ → 1/2, F0 → 1.2, Ω → 2/3},
      {ϕ, ω}, {t, 0, 900 π}, MaxSteps → 100 000] // Flatten
```

Out[43]= {ϕ → InterpolatingFunction[Domain: {{0., 2830.}} Output: scalar],

ω → InterpolatingFunction[Domain: {{0., 2830.}} Output: scalar]}

The phase diagram for the nonlinear pendulum when the driving force, $F_0 = 0.5$

```
In[44]:= ParametricPlot[Evaluate[{ϕ[t], ω[t]} /. phasePoints3],
      {t, 0, 90 π}, AxesLabel → {"ϕ", "ω"}, AspectRatio → 0.5]
```

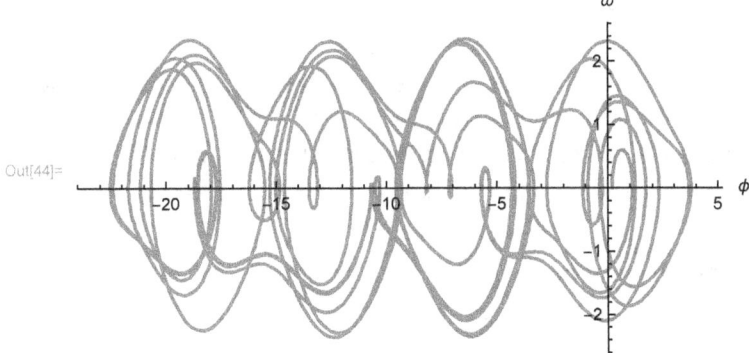

Figure 3.3.5. The phase diagram of a damped, driven nonlinear pendulum ($F_0 = 1.2$, $\gamma = 1/2$, $\Omega = 2/3$)

```
In[45]:= (* Plot the angle vs time, and the angular velocity vs time *)
Plot[{ϕ[t] /. phasePoints3[[1]], ω[t] /. phasePoints3[[2]]} // Evaluate,
 {t, 0, 90 π}, AxesLabel → {"ϕ", "ω"}, PlotRange → Full]
```

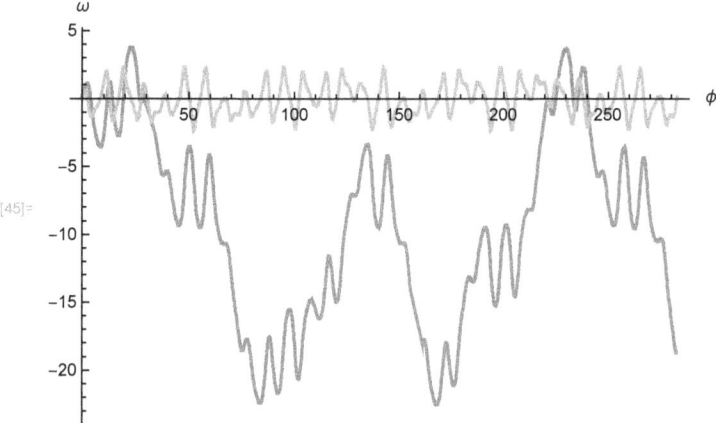

Out[45]=

Figure 3.3.6. The graph of ϕ and ω as a function of time for a damped, driven nonlinear pendulum (F_0 = 1.2, γ = 1/2, Ω = 2/3)

By increasing the amplitude of the driving force to F_0 = 1.2, the behavior of the system changes radically. The motion now is no longer simple. We might expect that for longer times, the motion might settle into some sort of repeating steady-state behavior. However, this is not the case, at least in the range shown here for this value of driving force. The behavior never repeats. This is an example of chaotic behavior.

At low driving force (F_0 = 0.5) the motion is a simple oscillation. After the transients have decayed, the motion would repeat forever at long times. On the other hand, at high driving force (F_0 = 1.2) the motion is chaotic. It is a very complicated nonrepeating function of time.

From the differential equations of motion with the initial conditions specified, we obtain the solution for ϕ as a function of time. The solution is then completely determined for all future times, but at the same time it can show random and unpredictable behavior for certain parameter (such as when we set the value F_0 = 1.2). We call this behavior **chaotic** when the motion exhibit deterministic and unpredictable behavior at the same time.

Recall that the solution we obtain from **NDSolve** command is in principle defined for any value of ϕ, that is, in the range of $-\infty$ to ∞, our pendulum can swing all way around its pivot point, which corresponds to $|\phi| > \pi$. Since ϕ is an angular variable, values of ϕ that differ by 2π in fact correspond to the same position of the pendulum. Therefore, for plotting purposes, we shall keep ϕ in the range $-\pi$ to π.

This is how we keep ϕ in range, $-\pi \leq \phi \leq \pi$. If ϕ becomes less than $-\pi$, then its value is increased by 2π, likewise, if it becomes greater than $+\pi$, its value is decreased by 2π.

We define a function to reduce the angle so as to keep it in the range of $-\pi$ to π.

```
In[46]:= reduced[ϕ_] := Mod[ϕ, 2 π] /; Mod[ϕ, 2 π] ≤ π

In[47]:= reduced[ϕ_] := (Mod[ϕ, 2 π] - 2 π) /; Mod[ϕ, 2 π] > π
```

Let us now apply the reduce function to our solutions.
For comparison, we plot the reduced ϕ function together with the original ϕ function on the same graph

```
In[48]:= (* Plot the reduced φ and the original φ vs time *)
Plot[Evaluate[{reduced[φ[t]], φ[t]} /. phasePoints3[[1]]],
 {t, 0, 30 π}, PlotStyle → {Automatic, {Red, Dashed}}, AxesLabel → {"φ", "t"}]
```

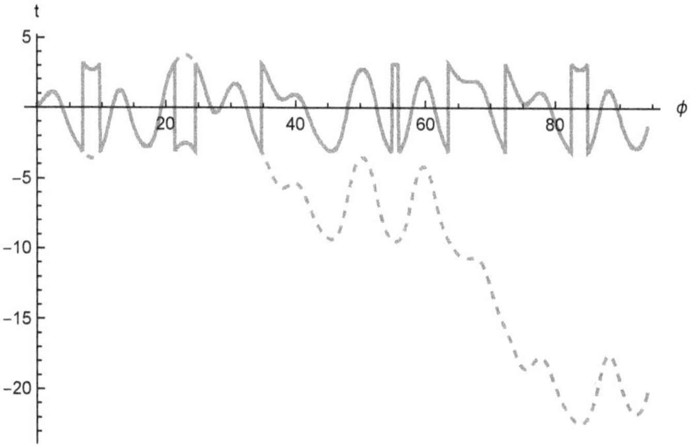

Figure 3.3.7. The behavior of the reduced ϕ and the original ϕ (dashed line) as a function of time. Notice that the vertical "jumps" in ϕ occur when the angle is reset so as to keep it in the range of $-\pi$ to π, they do not corresponds to the discontinuities in ϕ

```
In[49]:= (* Plot the reduced phase diagram of chaotic regime *)
reducedPhaseDiagram =
 ParametricPlot[Evaluate[{reduced[φ[t]], ω[t]} /. phasePoints3],
  {t, 0, 90 π}, AxesLabel → {"φ", "ω"}, PlotRange → All]
```

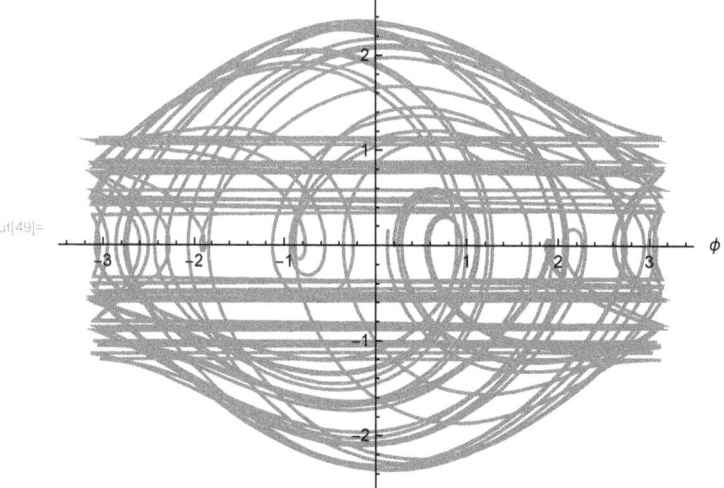

Figure 3.3.8. The reduced phase diagram of a damped, driven nonlinear pendulum. (F_0 = 1.2, γ = 1/2, Ω = 2/3) (it is in chaotic regime)

In[50]:= `(* Define a point on the phase diagram of chaotic regime *)`
`pointOnPhase[t_] := Graphics[`
`  {Red, PointSize[0.02], Point[{reduced[ϕ[t]], ω[t]} /. phasePoints3 /. t → 0]}]`

In[51]:= `Show[reducedPhaseDiagram, pointOnPhase[1]]`

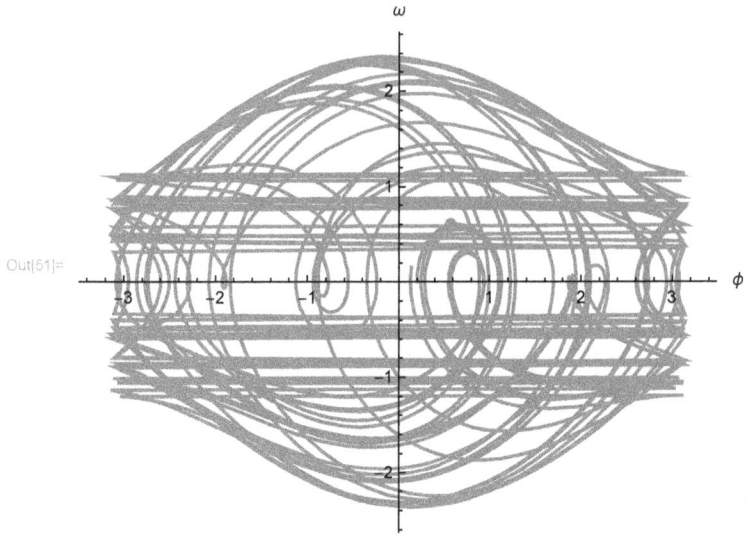

Here is an animation to show the movement of a particle in phase diagram

```
In[52]:= Manipulate[Show[reducedPhaseDiagram, pointOnPhase[t]],
        {{t, 0}, 0, 90 π, Appearance → {"Labeled"}}]
```

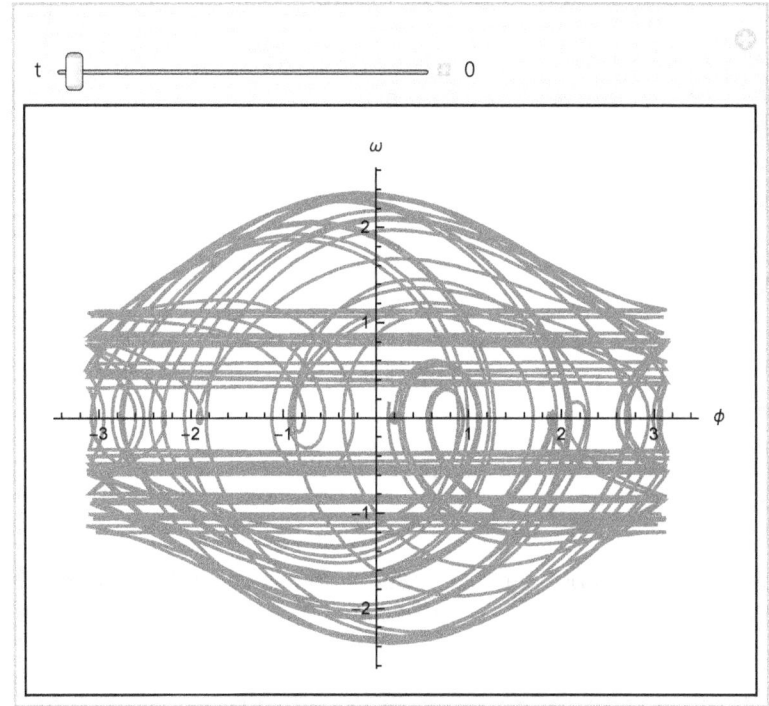

Figure 3.3.9. An animation of a moving particle on the phase diagram of a damped, driven nonlinear pendulum. ($F_0 = 1.2$, $\gamma = 1/2$, $\Omega = 2/3$)

The pattern of the phase diagram is not simple. The trajectory exhibits many orbits that are nearly closed and that persist for several cycles. Though it looks random, it exhibits some significant structures. This is a common property of chaotic systems.

Let us examine these trajectories in phase space in a slightly different manner. We graph the same phase diagram only at times that are in phase with the driving force. That is, we display only those points when $\Omega t = 2 n \pi$, where n is an integer.

```
In[53]:= ListPlot[
    Table[{reduced[ϕ[t]], ω[t]} /. phasePoints3, {t, 16, 900 π, (2 π / Ω) /. Ω → (2/3)}],
    PlotStyle → {Red, PointSize[0.01]}, AxesLabel → {"ϕ", "ω"}, PlotRange → All]
```

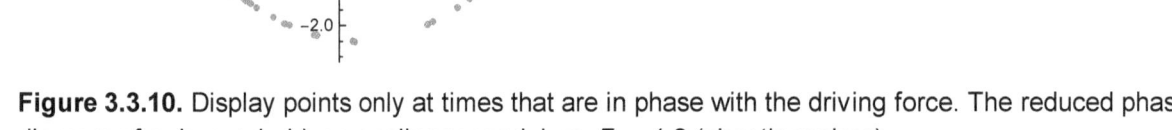

Figure 3.3.10. Display points only at times that are in phase with the driving force. The reduced phase diagram of a damped, driven nonlinear pendulum. F_0 = 1.2 (chaotic regime)

The result as shown in Figure 3.3.10 is an example of what is known as a **Poincaré section**. This surface of points is called a **strange attractor** of the nonlinear pendulum.

The Poincaré section represents a slice of the phase space of the system. For the three dimensional case, a slice can be obtained from the intersection of a continuous trajectory with a two dimensional plane in the phase space.

```
In[54]:= phaseSpace3D = ParametricPlot3D[
    Evaluate[{reduced[ϕ[t]], ω[t], reduced[Ω t]} /. (Ω → 2/3) /. phasePoints3],
    {t, 0, 90 π}, AxesLabel → {"ϕ", "ω", "Ω t"}]
```

Out[54]=

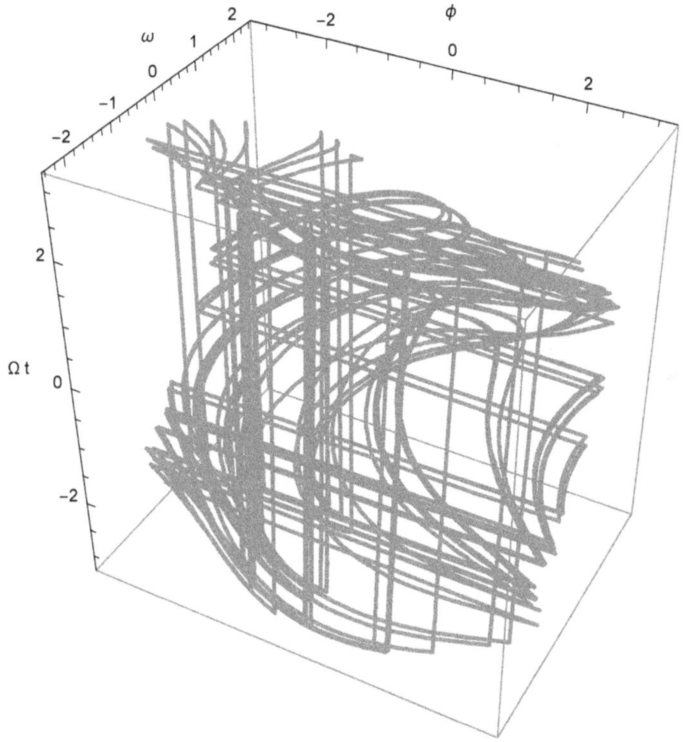

Figure 3.3.11. Three dimensional representation of the trajectory in phase space.

```
In[55]:= plane = ContourPlot3D[z == 2/3, {x, -4, 4}, {y, -4, 4},
    {z, -4, 4}, Mesh → None, ContourStyle → Directive[Opacity[0.8]]];
```

In[56]:= `Show[phaseSpace3D, plane]`

Out[56]=

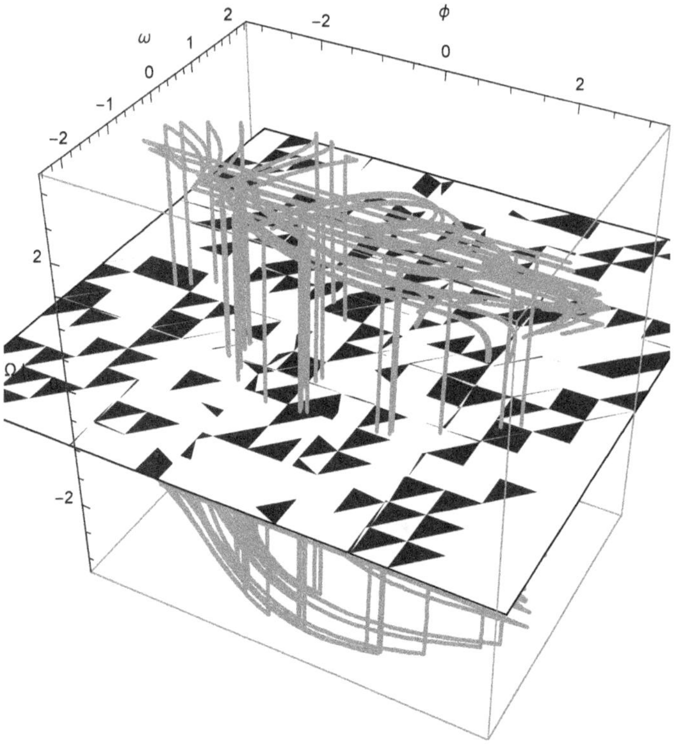

Figure 3.3.12. A Poincaré section represents a slice of the three-dimensional phase space of the system. For a damped, driven nonlinear pendulum, it is a plane of points which are in phase with the driving force

Thus, the Poincaré section represents a simpler view of the complicated chaotic trajectory and is a useful way to analyze the behavior of the dynamical system. It does so by recording the values of ϕ and ω at a rate that matches the characteristic frequency of the pendulum, which in present case the driving frequency Ω.

It is interesting to notice that except for the initial transient, the Poincaré section trajectory is the same for a wide range of initial conditions.

Exercise:
Change the initial conditions to $\phi(0) = 0.8$ and $\omega(0) = 0.5$, show that the Poincaré section is the same with the one in Figure 3.3.10.

Thus, even though $\phi(t)$ is unpredictable, we do know ϕ and ω are points on the phase space which reside on the Poincaré section. The trajectory of the pendulum is drawn to this section, thus, it is called **a strange attractor**.

In the case of non-chaotic regime ($F_0 = 0.5$), actually there is also attractor. Let us plot the attractor for such case,

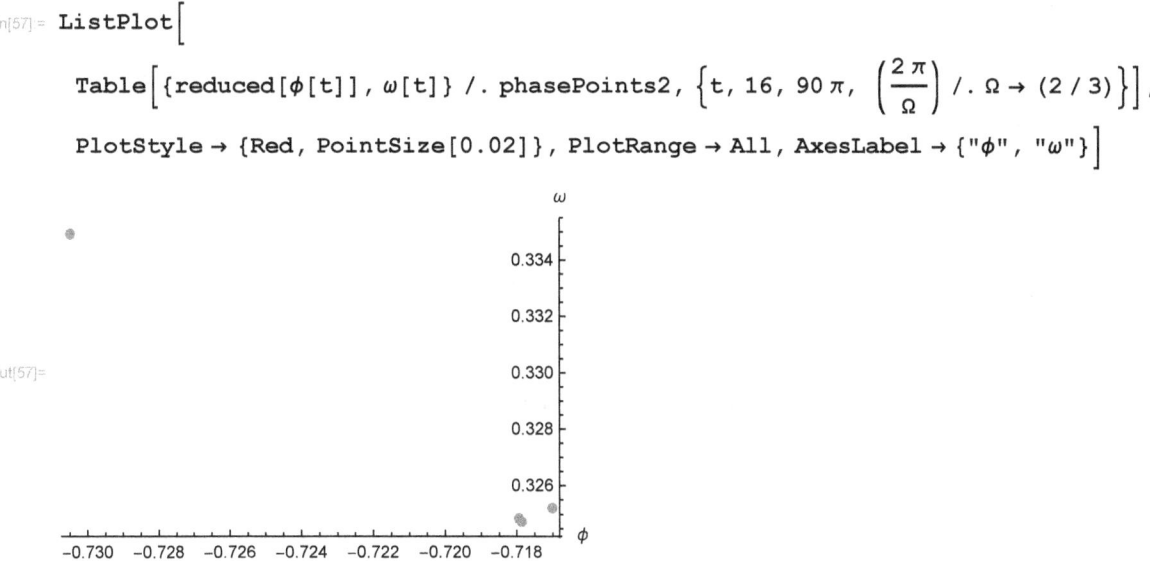

Figure 3.3.13. The reduced phase diagram of a damped, driven nonlinear pendulum, where $F_0 = 0.5$. The points are attractors in non-chaotic regime.

While the attractors have complicated structure in chaotic regime, they are just points in non-chaotic regime such as those shown in Figure 3.3.13.

We may then draw some conclusions from our discussions hitherto:

1. When a system is in chaotic regime, it can be both deterministic and unpredictable. However,

2. The behavior in chaotic regime is not completely random, but can be described by a Poincaré section in phase space.

3.3.2 Bifurcation

In previous section, we notice how the nonlinear pendulum responds to the driving force when the amplitude of the driving force is increased to $F_0 = 1.2$. We observed the chaotic behavior of the system in response to this new value. This raises question about the transition to chaotic regime of the system in response to changing of a certain parameter, in our case the amplitude of the driving force. A nice way to understand how the transition comes about is with a diagram which is known as a **bifurcation diagram**.

To construct a bifurcation diagram for the damped, driven nonlinear pendulum, we apply the following ways. For each value of F_0 we have calculated ϕ as a function of time. After bypassing 300 driving periods (for angular frequency $\Omega = 2/3$, this is 900π seconds) so that the initial transients have decayed away, we plotted ϕ at times that we were in phase with the driving force (just as we did in constructing the Poincaré section in Figure 3.3.10) as a function of F_0. Then we plot the points up to the 400th drive period. This process is then repeated for a range of values of F_0.

The following are the codes in Mathematica to implement this. The result is shown in Figure 3.3.14.

(**Notes:** The bifurcation diagram computations may take a few minutes, so be patient.)

```
In[58]:= phasePoints4[FD_] :=
    NDSolve[{eq331, eq332, ϕ[0] == 0.2, ω[0] == 0} /. {ω0 → 1, γ → 1/2, F0 → FD, Ω → 2/3},
     {ϕ, ω}, {t, 0, 1200 π}, MaxSteps → 100 000] // Flatten
```

```
In[59]:= ListPlot[Table[Table[Evaluate[{FD, reduced[ϕ[t]]} /. phasePoints4[FD]],
     {t, 900 π, 1200 π, (2π/Ω) /. Ω → (2/3)}], {FD, 1.420, 1.490, 0.0001}],
    PlotStyle → {Blue, PointSize[0.006]}, AxesLabel → {"F₀", "ϕ"},
    PlotRange → {{1.42, 1.49}, {0, 1.5}}]
```

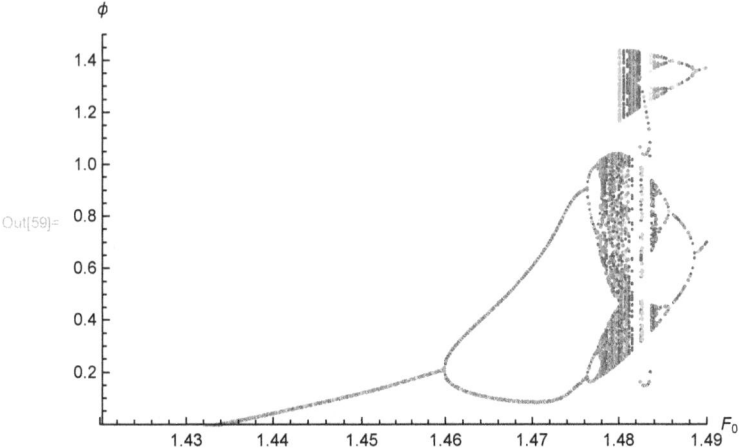

Figure 3.3.14. The bifurcation diagram for the damped, driven nonlinear pendulum

We notice from Figure 3.3.14, up to $F_0 = 1.46$, there is only one point of ϕ for each F_0, thus our bifurcation diagram consists of a single curve, although that point will be plotted many times. We refer to this as period-1 behavior.

As we increase the driving amplitude from 1.46 to about 1.476, then the values of ϕ that are plotted will alternate between two values, thus we see two curves in this range. We refer to this as period-2 behavior.

And so on with a pattern becomes obvious that for motion that is period-n will yield n points on the bifurcation diagram for that range of F_0. This period-doubling behavior leads the transition of the system to chaotic behavior.

3.4 The van der Pol oscillator

The van der Pol oscillator is a nonlinear damping oscillator, which was discovered by van der Pol (1926). The differential equation arises in the study of circuits containing vacuum tubes. It was also called relaxation oscillator and is often applied in the investigation of physical and biological phenomena such as electric circuits and models of the heartbeat, as well as the action potential in neurons.

3.4.1 Non-driven van der Pol Oscillator

The equation of motion for the non-driven van der Pol oscillator is

(3.4.1) $$\frac{d^2 x}{dt^2} - \gamma (A^2 - x^2) \frac{dx}{dt} + \omega_0^2 x = 0.$$

The second term on the left-hand side of the equation can either be an amplitude-dependent amplification or a frictional term. That is, if γ is a positive constant, it provides amplification for small amplitudes, but function as frictional factor for large amplitudes. Since energy is related to amplitudes, hence in such cases energy is fed into small oscillations and removed from large oscillations.

Let us consider a special case where $A = 1$, $\omega_0 = 1$, and the initial conditions be $(x_0, \dot{x}_0) = (5, 0)$.

We can solve this differential equation numerically using **NDSolve** command

```
In[60]:= Clear[x, y]
```

```
In[61]:= nsol[γvar_, x0_, v0_] :=
  NDSolve[{x''[t] - γ (A^2 - x[t]^2) x'[t] + ω0^2 x[t] == 0, x[0] == x0, x'[0] == v0} /.
    {A → 1, γ → γvar, ω0 → 1}, x, {t, 0, 50 π}]
```

```
In[62]:= (* For large amplitude, example here: x0=5,
  γ=0.05 functions as a frictional factor
  *)
Plot[Evaluate[x[t] /. nsol[0.05, 5, 0]],
  {t, 0, 50 π}, AxesLabel → {"t", "x"}, PlotRange → All]
```

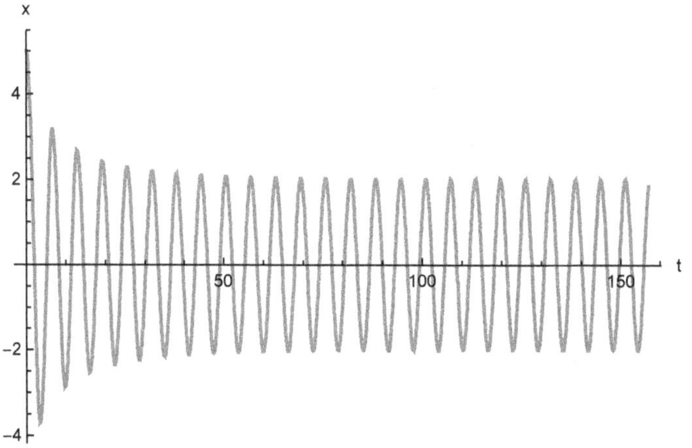

Figure 3.4.1.1. For large amplitude $x_0 = 5$, γ functions as a frictional term ($A = 1$, $\omega_0 = 1$, $\gamma = 0.05$)

Chapter 3. Nonlinear Oscillatory Systems

```
In[63]:= (* For small amplitude, example: 0.5,
    γ=0.05 provides amplification for the displacement *)
    Plot[Evaluate[x[t] /. nsol[0.05, 0.5, 0]], {t, 0, 50 π}, AxesLabel → {"t", "x"}]
```

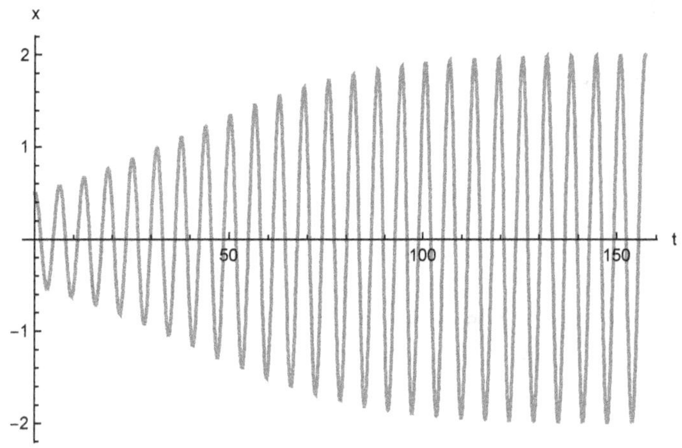

Figure 3.4.1.2. For small amplitude $x_0 = 0.5$. The oscillation grows at first, and as the amplitude become larger, it then saturated due to friction ($A = 1$, $\omega_0 = 1$, $\gamma = 0.05$)

The second-order differential equation of van der Pol oscillator (equation 3.4.1) can be reduced to two first-order differential equations by setting $y = \dot{x}$ as follows:

$$\frac{dx}{dt} = y,$$

$$\frac{dy}{dt} = \gamma\left(A^2 - x^2\right) y - \omega_0^2\, x.$$

By considering a special case where $A = 1$, $\omega_0 = 1$, we now solve this set of equations numerically using **NDSolve** command as follows

```
In[64]:= Clear[x, y]
```

```
In[65]:= nsol[γvar_, x0_, y0_] := NDSolve[
    Evaluate[{x'[t] == y[t], y'[t] == γ (A^2 - x[t]^2) y[t] - ω0^2 x[t], x[0] == x0, y[0] == y0} /.
    {A → 1, γ → γvar, ω0 → 1}], {x, y}, {t, 0, 50 π}]
```

In[66]:= `(* The graph of solutions of x and y in time *)`
`Plot[Evaluate[{x[t], y[t]} /. nsol[1, -3, 3]], {t, 0, 10 π}]`

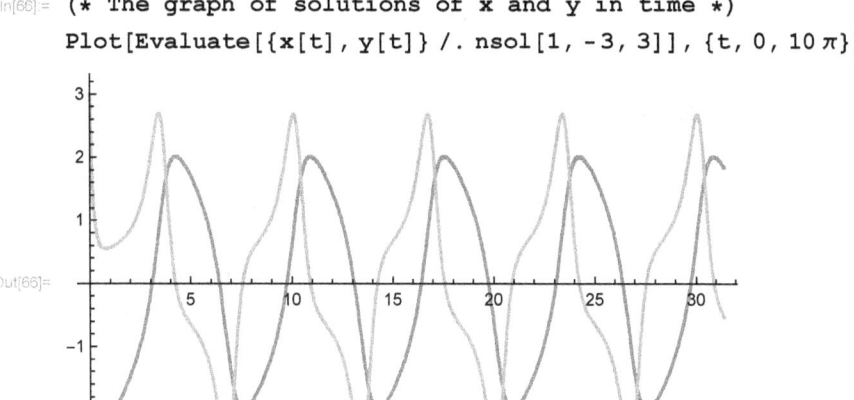

Figure 3.4.1.3. Time variations in *x* and *y* in the van der Pol oscillator. $\gamma = 1$. Initial conditions $(x_0, y_0) = (-3, 3)$

In[67]:= `(* Phase space (x vs y) plot with γ set to 1 for initial conditions (x₀,y₀)=(-3,3) *)`
`ParametricPlot[Evaluate[{x[t], y[t]} /. nsol[1, -3, 3]], {t, 0, 50 π}, ImageSize → 300]`

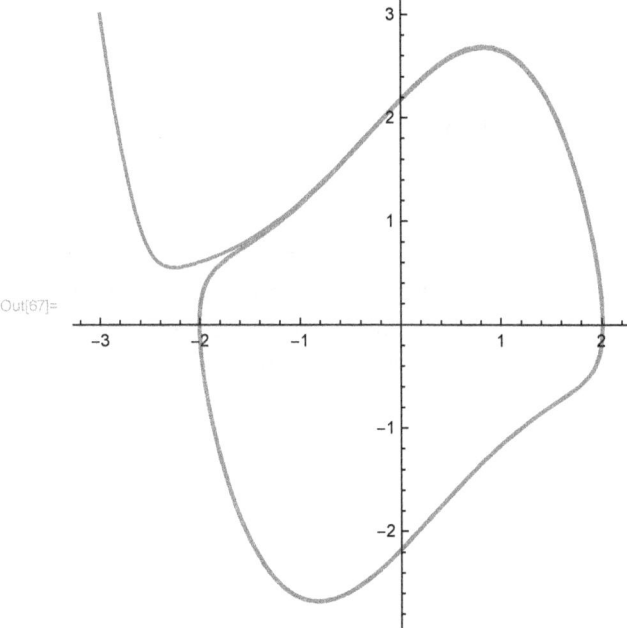

Figure 3.4.1.4. Phase space plot (*x* vs *y*) for the van der Pol oscillator. $\gamma = 1$. Initial conditions $(x_0, y_0) = (-3, 3)$

```
In[68]:= (* Phase space (x vs y) plot with γ set to 1 for various initial conditions
        *)
        toshow = ParametricPlot[Evaluate[{x[t], y[t]} /. #],
            {t, 0, 50 π}, AspectRatio → 1, PlotRange → All] & /@
```
$\left\{\text{nsol}[1, -3, 3], \text{nsol}[1, -2, 2], \text{nsol}[1, -1, 1], \text{nsol}\left[1, -\frac{1}{2}, \frac{1}{2}\right]\right\};$

We draw the phase diagrams of $x(t)$ versus $y(t)$ using **ParametricPlot** command and make it as a function by using the **&** command. The parameter represented by **#** symbol is applied with **nsol** by using **Map** command (**/@**) for various initial conditions ($(x_0, \dot{x}_0) = (-3, 3)$, $(x_0, \dot{x}_0) = (-2, 2)$, $(x_0, \dot{x}_0) = (-1, 1)$, $(x_0, \dot{x}_0) = (-1/2, 1/2)$). Finally, the graphs are assigned to **toshow**.

To draw the phase diagrams, we use **GraphicsGrid** command and partition its length to 4

```
In[69]:= GraphicsGrid[Partition[toshow, 4], ImageSize → 600]
```

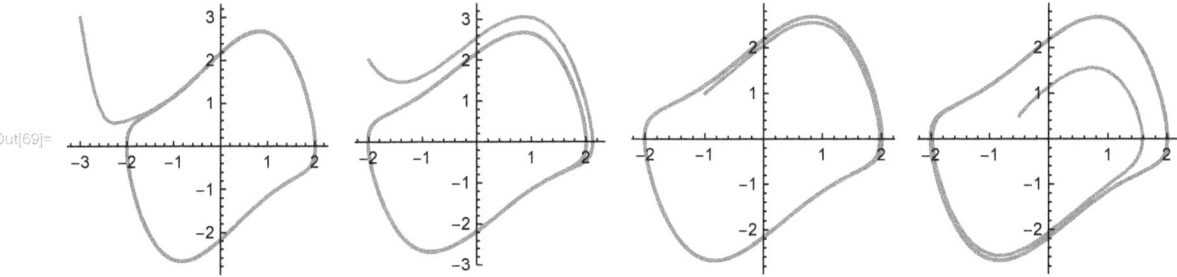

Figure 3.4.1.5. Limit-cycle orbits in phase space for the van der Pol oscillator expressed with various initial conditions.

In Figure 3.4.1.5 we show the phase space (**x** vs **y**) plot for $y = 1$ with various initial conditions. Initially, the trajectories vary with difference initial values. However, they later merge into a single circuit and rotate along the circuit. Thus, the final circuit does not depend on the initial conditions. This type of orbit is called a **limit cycle**. However, it can be verified that for $y < 0$, the trajectory does not form a limit cycle.

3.4.2 Driven van der Pol Oscillator

Suppose the van der Pol oscillator is driven by a sinusoidal force such that the equation of motion becomes

(3.4.2.1) $$\frac{d^2 x}{dt^2} - \gamma \left(A^2 - x^2\right) \frac{dx}{dt} + \omega_0^2 x = \alpha \cos \omega t,$$

where α is the amplitude of the driving force and ω is the driving frequency.

Let us consider a special case where $A = 1$, $\omega_0 = 1$, so the equation of motion becomes

(3.4.2.2) $$\frac{d^2 x}{dt^2} - \gamma(1-x^2)\frac{dx}{dt} + x = \alpha \cos \omega t.$$

Let us also define the following terms
$$x \equiv x,$$
$$y \equiv \frac{dx}{dt},$$
$$z \equiv \omega t.$$

Then, the equation of motion can be written as a set of three first-order differential equations as follows

(3.4.2.3) $$\frac{dx}{dt} = y,$$
$$\frac{dy}{dt} - \gamma(1-x^2)y + x = \alpha \cos z,$$
$$\frac{dz}{dt} = \omega.$$

We solve this set of equations numerically by assuming $(x_0, y_0, z_0) = (0, 0, 0)$.

```
In[70]:= Clear[x, y, z]
```

```
In[71]:= (* Equations of motion (3.4.2.3) *)
eq = {y'[t] - γ (1 - x[t]^2) y[t] + x[t] == α Cos[z[t]], x'[t] == y[t], z'[t] == ω}
Out[71]= {x[t] - γ (1 - x[t]^2) y[t] + y'[t] == α Cos[z[t]], x'[t] == y[t], z'[t] == ω}
```

```
In[72]:= initConditions = {x[0] == 0, y[0] == 0, z[0] == 0}
Out[72]= {x[0] == 0, y[0] == 0, z[0] == 0}
```

Case 1: Consider the following conditions:
$$\alpha = 0.1, \quad \gamma = 0.05, \quad \omega = 1.$$

Chapter 3. Nonlinear Oscillatory Systems

```
In[73]:= numsol = NDSolve[
          Evaluate[{eq, initConditions} /. {α → 0.1, γ → 0.05, ω → 1}], {x, y, z}, {t, 0, 200 π}]

Out[73]= {{x → InterpolatingFunction[ Domain: {{0., 628.}} Output: scalar ],

          y → InterpolatingFunction[ Domain: {{0., 628.}} Output: scalar ],

          z → InterpolatingFunction[ Domain: {{0., 628.}} Output: scalar ]}}
```

```
In[74]:= (* The phase phase plot: x vs y *)
         ParametricPlot[{x[t], y[t]} /. numsol // Evaluate, {t, 0, 20 π}, AxesLabel → {"x", "y"}]
```

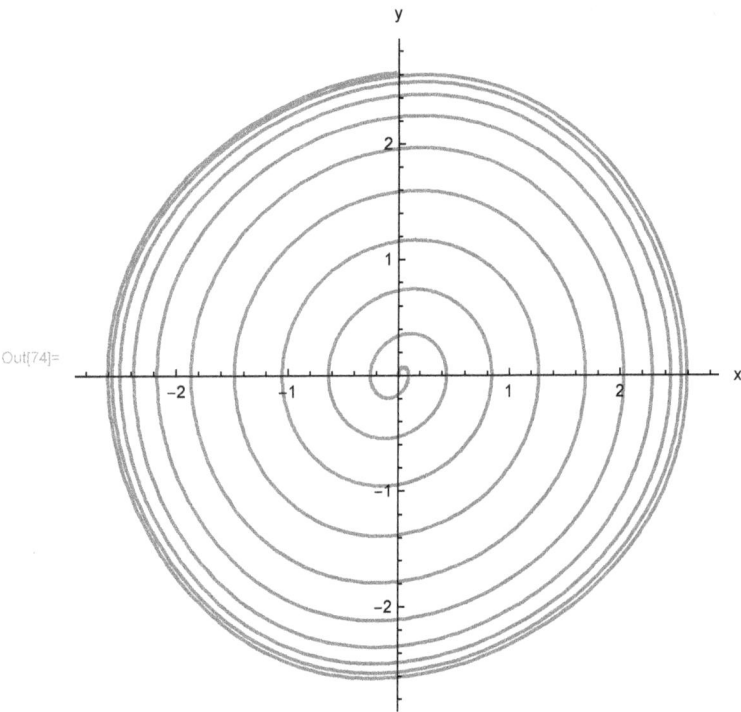

Figure 3.4.2.1. The phase phase plot x vs y ($\alpha = 0.1$, $\gamma = 0.05$, $\omega = 1$).
The system is in **non-chaotic** regime.

```
In[75]:= (* Plot the position x(t) vs time *)
       Plot[x[t] /. numsol // Evaluate, {t, 0, 50 π}, AxesLabel → {"t", "x"}]
```

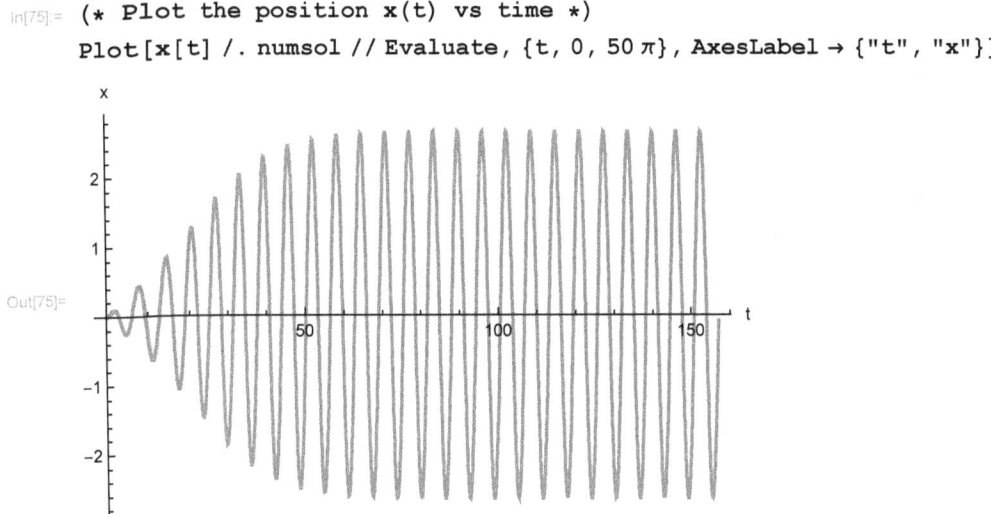

Figure 3.4.2.2 The position x(t) as a function of time (**non-chaotic regime**)

```
In[76]:= (* The phase space plot *)
       ParametricPlot3D[{x[t], y[t], z[t]} /. numsol,
        {t, 0, 20 π}, PlotRange → {{-3, 3}, {-3, 3}, {0, 20 π}},
        AspectRatio → 2, PlotStyle → AbsoluteThickness[4]]
```

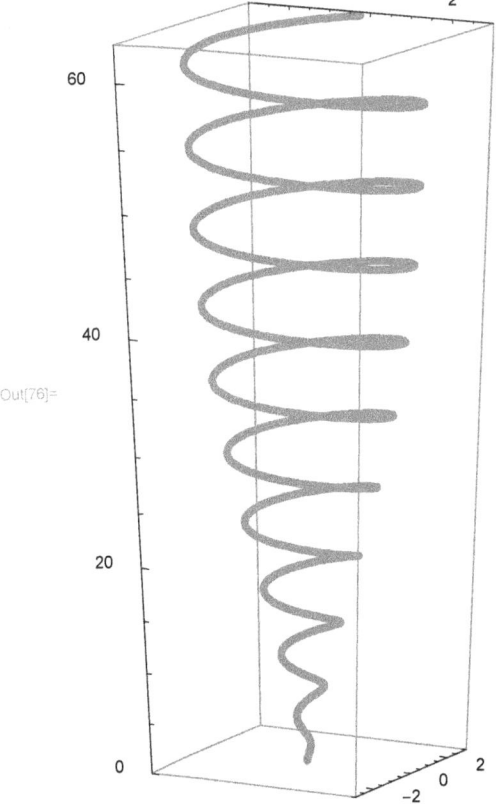

Figure 3.4.2.3 The three-dimensional phase-space plot of the motion (**non-chaotic regime**)

Chapter 3. Nonlinear Oscillatory Systems

Case 2: Consider the following conditions:

$$\alpha = 5, \quad \gamma = 5, \quad \omega = 2.466.$$

In[77]:= `numsol = NDSolve[Evaluate[{eq, initConditions} /. {α → 5, γ → 5, ω → 2.466}],`
`{x, y, z}, {t, 0, 50 π}, MaxSteps → 100 000]`

Out[77]= {{x → InterpolatingFunction[... Domain: {{0., 157.}} Output: scalar],

y → InterpolatingFunction[... Domain: {{0., 157.}} Output: scalar],

z → InterpolatingFunction[... Domain: {{0., 157.}} Output: scalar]}}

In[78]:= `(* The phase plane plot: x vs y *)`
`ParametricPlot[{x[t], y[t]} /. numsol, {t, 0, 50 π}, AxesLabel → {"x", "y"}]`

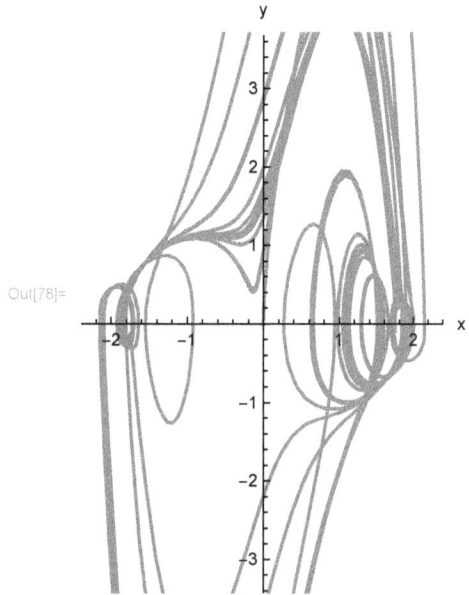

Figure 3.4.2.4. The phase plane plot x vs y ($\alpha = 5$, $\gamma = 5$, $\omega = 2.466$). The system is in **chaotic** regime.

```
In[79]:= (* Plot the position x(t) vs time *)
       Plot[x[t] /. numsol, {t, 0, 50 π}, AxesLabel → {"t", "x"}]
```

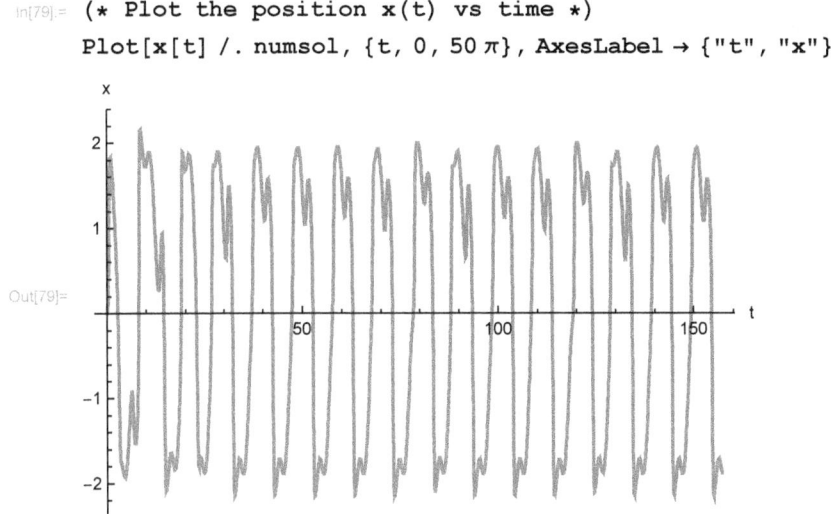

Figure 3.4.2.5 The position $x(t)$ as a function of time (**chaotic regime**)

We notice that the van der Pol oscillator exhibits chaotic behavior when the values of these parameters becomes $\alpha = 5$, $\gamma = 5$, $\omega = 2.466$.

```
In[80]:= (* The phase space plot *)
       ParametricPlot3D[{x[t], y[t], z[t]} /. numsol,
        {t, 0, 20 π}, PlotRange → {{-3, 3}, {-3, 3}, {0, 50 π}},
        PlotStyle → AbsoluteThickness[4], AxesLabel → {"x", "y", "z"}, AspectRatio → 2]
```

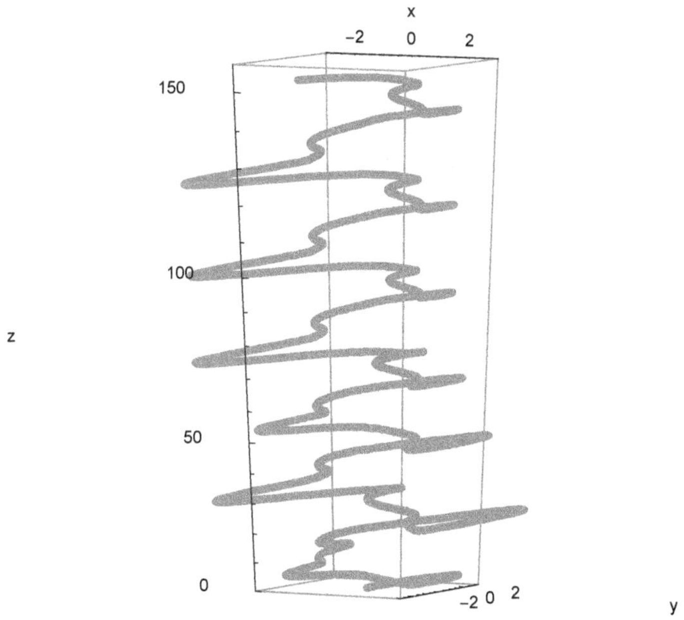

Figure 3.4.2.6 The three-dimensional phase-space plot of the motion (**chaotic regime**)

This page intentionally left blank

Bibliography

Abell, M. L., Braselton, J. P., Mathematica by Example 4th ed., Elsevier Inc., 2009

Andronov, A. A., Vitt, A. A., Khaikin, S. E., Theory of Oscillators, Pergamon Press Ltd. 1966

Baumann, G., Mathematica for Theoretical Physics, Classical Mechanics and Nonlinear Dynamics, 2nd ed., Springer Science + Business Media, Inc., New York, 2005

Don, E., Schaum's Outlines Mathematica 2nd ed., The McGraw-Hill Companies, Inc., 2009

Fowles G. R. , Cassiday G. L., Analytical Mechanics, 7th ed. Thomson Brooks/Cole, California, 2005

Giordano, N. J., Nakanishi, H., Computational Physics 2nd ed., Pearson Education, Inc., New Jersey, 2006

Kinoshita, S. (editor), Pattern Formations and Oscillatory Phenomena, Elsevier Inc., 2013

Kneubuhl, F. K., Oscillations and Waves, Springer-Verlag Berlin Heidelberg, New York, 1997

Thornton, S. T., Marion, J. B., Classical Dynamics of Particles and Systems 5th ed., Thomson Brooks/Cole, California, 2004

Torrence, B. F., Torrence, E. A., The Student's Introduction to Mathematica®–A Handbook for Precalculus, Calculus, and Linear Algebra 2nd ed. Cambridge University Press, New York, 2009

Weisstein, Eric W. "Lissajous Curve." From MathWorld--A Wolfram Web Resource. http://mathworld.wolfram.com/LissajousCurve.html

Weisstein, Eric W. "van der Pol Equation." From MathWorld--A Wolfram Web Resource. http://mathworld.wolfram.com/vanderPolEquation.html

www.ingramcontent.com/pod-product-compliance
Lightning Source LLC
Chambersburg PA
CBHW080930170526
45158CB00008B/2232